Finite Element Analysis of Non-Newtonian Flow

Springer-Verlag London Ltd.

Hou-Cheng Huang, Zheng-Hua Li and Asif S. Usmani

Finite Element Analysis of Non-Newtonian Flow

Theory and Software

With 31 Figures

 Springer

Dr Hou-Cheng Huang, PhD, CEng, MIMechE, NRA
5 Segsbury Road, Wantage, Oxfordshire OX12 9XR, UK

Dr Zheng-Hua Li, PhD
7 Dorcas Avenue, Stoke Gifford, Bristol BS34 8XG, UK

Dr Asif S. Usmani, BE, MS, PhD
29 Burdiehouse Road, Edinburgh EH9 3JN, UK

Additional material to this book can be downloaded from http://extras.springer.com.

ISBN 978-1-4471-1204-4

British Library Cataloguing in Publication Data
Huang, Hou-Cheng, 1947-
  Finite element analysis of non-Newtonian flow : theory and software
  1. Non-Newtonian fluids 2. Finite element method
  3. Non-Newtonian fluids - Mathematics 4. Non-Newtonian fluids
  - Computer programs
  I. Title II. Li, Zheng Hua III. Usmani, Asif S.
  532'.051
ISBN 978-1-4471-1204-4

Library of Congress Cataloging-in-Publication Data
Huang, Hou-Cheng, 1947-
  Finite element analysis of non-Newtonian flow : theory and software /
  Hou-Cheng Huang, Zheng-Hua Li and Asif S. Usmani.
      p.   cm.
  Includes bibliographical references and index.

  ISBN 978-1-4471-1204-4   ISBN 978-1-4471-0799-6 (eBook)
  DOI 10.1007/978-1-4471-0799-6

  1. Non-Newtonian fluids 2. Finite element method. 3. Finite element method--Software.
  I. Li, Zheng Hua, 1952-   II. Usmani, Asif S. (Asif Sohail), 1959-   III. Title.
QC189.5.H8   1998   98-14488
620.1'06--dc21   CIP Rev.

Typesetting: Camera ready by authors
Printed and bound at the Athenæum Press Ltd., Gateshead, Tyne & Wear
69/3830-543210 Printed on acid-free paper

To our families

Preface

Today, the finite element method has been widely employed in solving field problems arising in modern industrial practices. The text presented here is an introduction to the application of the finite element method to the analysis of non-Newtonian fluid flow which is a very common phenomenon in many processes of manufacturing, such as polymer flow, extrusion, spinning, injection and blow moulding.

The discussion has been limited to laminar behaviour of non-Newtonian flow for both steady state and transient analysis. The book contains the basic fluid flow concepts and some typical constitutive equations for the viscous shear flow including both elastic and inelastic liquids. In the computational implementation, we have also presented some modern techniques which are being used to enhance the accuracy and speed of the conventional method. In writing the text we have endeavoured to keep it accessible to persons with qualifications of no more than an engineering graduate. As mentioned earlier this book may be used to learn FEM by beginners, this may include undergraduate students and practicing engineers.

The first three chapters include a review of the basic non-Newtonian concepts, the governing equations and a gradual introduction to the Finite Element Method with CFD concepts. Chapter 4 focuses on Steady non-Newtonian Flow with a mixed formulation. The last two chapters contribute to transient analysis and adaptive techniques.

Finally two appendices give complete details of the programs NSTEAD and NFLOWG, with full user instructions and documented examples. NSTEAD is a program for steady non-Newtonian problems whereas NFLOWG is for transient analysis. The complete source code for both the programs and sample input data files are provided on a floppy disc included with the book.

Hou-Cheng Huang, Zheng Hua Li, Asif S. Usmani

1998

Conditions for Program Usage

1. The CD-ROM may only be used by the book purchaser on his/her microcomputer.

2. A back-up copy of the CD-ROM may be made by the book purchaser.

3. The program may not be transferred onto any other machine without the written permission of the author, or in the case of his death, by the written permission of his estate.

4. The software is supplied without any warranties of any kind, whether implied or otherwise.

5. All queries regarding function, performance or adjustments to the programs should be directed to the authors at the addresses shown on page iv.

Additional Note

The print-outs of the CD-ROM's contents shown in the Appendices of this book have been especially formatted for reproduction; users should not expect their print-outs to match those in this section.

Contents

Chapter 1

Introduction

1.1 Preamble

A great deal of computational research has been undertaken and published in the field of Computational Fluid Dynamics (CFD) since the advent of the digital computer. Before 1970, the Finite Difference Method (FDM) was almost universally used as computer based numerical method in modelling fluid dynamical process [1]. Since then there has been a revolution in the general area of mathematical modelling. Highly sophisticated and detailed analysis of many engineering problems has become possible. However, it can be argued that the last three decades have in many ways belonged to the Finite Element Method (FEM) as the method of choice among the currently available numerical methods for solving mathematical equations [2]. This is borne out by the fact that FEM has been applied to as vast an array of physical problems as one can possibly imagine. Not surprisingly, fluid dynamics being one of the oldest branches of physics, has consequently been one of the main arenas of activity for researchers and practitioners of FEM. Despite the continued use of FDM and related techniques for routine fluid dynamics problems, FEM is increasingly the preferred numerical method for the analysis of the most complex types of flow problems with unrivalled accuracy.

The vast majority of CFD related research has concentrated on compressible or incompressible Newtonian fluids [3], [4]. Such fluids have a constant viscosity which is independent of the velocity gradient, or stated another way, the stress in Newtonian fluids is proportional to the rate of shear. There exist however, a fairly large category of liquids for which the viscosity is not independent of the rate of shear and these liquids are referred to as non-Newtonian. Exact solutions for non-Newtonian flows are practically impossible. This necessitates the use of numerical methods for obtaining approximate solutions to most non-Newtonian flow problems. This book will concentrate on the application of FEM to non-Newtonian fluid flow.

There are a number of FE techniques which can be used to model steady and transient non-Newtonian flow [5]. In this book we will use techniques based only on the *primitive variable* formulations (*i.e.* velocity components and pressure) of the flow equations. There are two main classes of methods using such formulations, namely, *mixed* and *segregated* methods. Both of these methods have been used in this book, mixed methods for steady non-Newtonian flow (Chapter 4) and segregated methods for transient (Chapter 6). The details of background theory and implementation of both these types of methods appear in the relevant chapters. For transient analysis, an adaptive remeshing technique has also been introduced in Chapter 6.

FEM is an approximate numerical method and care has to be exercised in setting up a problem for FEM analysis. The quality of the solution obtained depends upon various factors including mainly the distribution of the space discretisation (meshing) throughout the domain, time discretisation for transient problems, proper application of the boundary conditions and selection of suitable material properties. All of these aspects of setting up a problem require diligence and experience. Given that proper care has been exercised in setting up the problem for analysis, the results are generally very reliable and provide valuable insight to the designer. However, one needs to be sceptical of all results as errors can creep into the analysis from unforeseeable sources. Therefore it is essential to examine the results carefully and look for anomalies or inconsistencies in them using engineering intuition. The high quality graphical visualisation of results available to the modern user can be of benefit in the interpretation and examination of results. This can also be a source of false security for an unsuspecting user. Therefore it is important to maintain a healthy scepticism and a critical outlook towards the results obtained from all numerical methods.

Without reference to any particular application, this text provides means to the reader/user to enable him to solve different non-Newtonian flow problems, some of which may be listed as follows:

1. Steady State Analysis:

   - Creep motion
   - Slippage and friction
   - Power law and plastic flow

2. Transient Analysis:

   - Taylor-Galerkin technique
   - Projection method
   - Pressure correction scheme

3. Adaptive remeshing technique

1.2 Objectives and Layout

The main objectives of this book may be listed as follows:

1. Remind the reader of the basic equations that govern various forms of fluid flow.

2. Explain the background and fundamentals of the finite element method.

3. Describe the general procedure of achieving spatial and temporal discretisation of the governing equations via the finite element method.

4. Introduce special techniques for dealing with time integration.

5. Extend the basic FEM approach to include adaptive analysis based on error estimation.

6. Provide a fully documented set of software with the source code to the reader with test examples covering most topics in the text.

The text in the following chapters has been arranged according to the objectives stated above. The next chapter has been devoted to the establishment of the basic differential equations that govern general fluid dynamical problems. The constitutive laws for non-Newtonian fluids are discussed as a special case in which the constitutive equations can be degenerated to:

$$\boldsymbol{\tau} = \mu \frac{\partial u}{\partial s} \tag{1.1}$$

where $\boldsymbol{\tau}$ is the shear stress and μ is the coefficient of viscosity of the fluid (often referred to as just viscosity) and $\frac{\partial u}{\partial s}$ is the velocity gradient. Equation 1.1 is Newton's hypothesis that states the stress in a Newtonian fluid is directly proportional to the velocity gradient.

Chapter 3 introduces the reader to the mathematical background of the finite element method, following which the basic concepts are explained. The spatial discretisation of the governing differential equations is then demonstrated using the basic principles established earlier.

Chapter 4 describes the application of FEM to solve steady state flow problems, especially the creeping motion problem, where either the viscosity is very large or the velocity exceedingly small. As mentioned earlier, a *mixed* or *integrated* formulations of the steady state Navier-Stokes equations is used in this chapter.

Discretisation in the time domain for transient problems is discussed in Chapter 5. Various schemes that are commonly used for transient analyses are presented.

Special numerical techniques used for solving transient non-Newtonian flow are described in Chapter 6. These techniques include the Taylor-Galerkin method and the pressure correction method, which is a variant of the various *segregated* formulations of the Navier-Stokes equations which have become increasingly prevalent over the last decade.

To obtain an accurate solution, it is sometimes necessary to use very fine meshes in regions of a problem where high gradients of the field variable exist. The variation of field variable in the rest of the domain may, on the contrary, be so gentle as to require a very coarse mesh for its adequate resolution. The areas of high gradient may not always be predictable and in the case of transient problems, they may not be restricted to a particular region. Therefore if a fixed fine mesh is used in solving such problems, it may turn out to be unacceptably expensive. For such problems, the technique of adaptive analysis based on estimating the discretisation error provides an economical alternative. This technique is also discussed for flow problems in Chapter 6.

The Appendices describe the software included with the text. The main variables used are defined and some key subroutines are listed. The program structures are illustrated schematically. Full user instructions are given for the two programs NSTEAD and NFLOW. Several documented examples are also included for both the programs to familiarise the user with the software. Appendix A presents the program NSTEAD, which may be used for steady non-Newtonian flow problems with power law constitutive equations. This software can also be used to simulate metal forming processes.

Appendix B introduces the program NFLOW, which is for transient non-Newtonian flow problems. The Taylor-Galerkin Pressure correction method is implemented in NFLOW.

Bibliography

[1] B.P.Leonard. A survey of finite differences of opinion on numerical muddling of the incomprehensible defective confusion equation. In T.J.R.Hughes, editor, *Finite Element Methods for Convection Dominated Flows*, ASME, AMD, 1979.

[2] O.C.Zienkiewicz. *The Finite Element Method*. McGraw-Hill Book Company (UK) Limited, London, 1977.

[3] R.Peyret and T.D.Taylor. *Computational Methods for Fluid Flow*. Springer-Verlag, New York, USA, 1983.

[4] F.Thomasset. *Implementation of Finite Element Methods for Navier-Stokes Equations*. Springer-Verlag, New York, USA, 1981.

[5] M.J.Crochet. *Numerical Simulation of Non-Newtonian Flow*. Elsevier, New York, U.K., 1984.

Chapter 2

Governing Differential Equations

2.1   Introduction

The governing equations in non-Newtonian fluid mechanics are derived from the principles of *mechanics* which can be applied indiscriminately to all materials and the science of *rheology* which is the study of *deformation* and *flow* behaviour of real materials. These sciences are the foundation of many engineering disciplines such from strength of materials and structural engineering to hydraulics and fluid mechanics. However, due to the very broad class of material behaviour that may be represented within the classification of non-Newtonian *fluids*, the study of non-Newtonian fluids is often simply called *rheology* [1,2].

In the isothermal case, the principles of mechanics give rise to *field* equations. These are the equations of continuity, derived from the mass conservation principle and the stress equations of motion which are also called the equilibrium equations, derived from the application of Newton's second law of motion (conservation of momentum) to a fluid volume. The type of material determines the *constitutive* equations which relate the stresses in the material to the velocity of motion. Therefore, the governing equations of non-Newtonian flow include:

- The continuity equation

- The momentum equations

- The constitutive equations.

2.2   Governing Equations

In deriving the governing equations of fluid flow, the standard Eulerian approach using a cartesian coordinate system is to consider a control volume. Then for each quantity (*mass* and *momentum*), the conservation principle is invoked *i.e.*, the rate of storage of the quantity in question in the control volume, is equal to the rate of inflow minus the rate of outflow plus the rate of production within the volume. For momentum (after appropriate simplifications applicable here) this leads to,

$$\rho \frac{D\mathbf{v}}{Dt} = \rho \mathbf{g} + \nabla \cdot \sigma \tag{2.1}$$

where

$$\frac{D}{Dt} = \frac{\partial}{\partial t} + \mathbf{v} \cdot \nabla$$

and the $\mathbf{v}$ is the fluid velocity vector with components u, v and w in x, y and z directions, respectively, and t represents time, ρ the mass density and $\mathbf{g}$ the gravity.

If the hydraulic pressure term is separated from σ in the equation (2.1) as follows

$$\sigma_{ij} = -P\delta_{ij} + \tau_{ij}$$

then, the equation (2.1) becomes

$$\rho \frac{D\mathbf{v}}{Dt} = \rho \mathbf{g} + \nabla \cdot \boldsymbol{\tau} - \nabla P \tag{2.2}$$

where P is the pressure and $\boldsymbol{\tau}$ is commonly known as the *extra stress tensor* which is related to deformation rate by the constitutive equation. Equations (2.1) and (2.2) are also known as the Navier-Stokes equations.

Equating the quantity of mass entering and leaving the elemental volume, we obtain the conservation equation in vector notation as follows,

$$\frac{\partial \rho}{\partial t} + \nabla \cdot (\rho \mathbf{v}) = 0 \tag{2.3}$$

or

$$\frac{\partial \rho}{\partial t} + \frac{\partial}{\partial x}(\rho u) + \frac{\partial}{\partial y}(\rho v) + \frac{\partial}{\partial z}(\rho w) = 0 \qquad (2.4)$$

If the density for the problems relevant to us is considered to be constant, *i.e.* incompressible material, the above equations reduce to a statement of continuity

$$\nabla \cdot \mathbf{v} = 0 \qquad (2.5)$$

or

$$\frac{\partial u}{\partial x} + \frac{\partial v}{\partial y} + \frac{\partial w}{\partial z} = 0. \qquad (2.6)$$

2.3 Constitutive Equation

Examining Equations (2.1) and (2.2), an analogy can be observed in the behaviour of an elastic solid and a viscous fluid. With P analogous to normal stress components and τ analogous to the *deviatoric* components of the stress tensor σ. It is these deviatoric or *extra* stresses that are related to the deformation rate (or the rate of strain). In the following section we describe the relationship between the extra stresses and the deformation rate.

2.3.1 Viscous shear flow

That a viscous fluid in motion sustains shear stresses, has been observed from numerous experiments. In general, the relationship between the shear stresses and the deformation rate can be written as

$$\tau_{ij} = \alpha \delta_{ij} + \beta d_{ij} + \gamma d_{ik} d_{kj} \qquad (2.7)$$

where α, β and γ are function of three scalar invariants of d_{ij}, namely

$$I_1 = d_{ii}$$

$$I_2 = \frac{1}{2} d_{ij} d_{ij}$$

$$I_3 = \frac{1}{3} d_{ij} d_{jk} d_{ki}$$

The deformation rate tensor **d** is defined as

$$\mathbf{d} = \frac{1}{2}(u_{i,j} + u_{j,i}) \tag{2.8}$$

where, $d_{i,j}$ denotes elements of the tensor **d**, i and j are taken as 1, 2 and 3 and repeated indices indicate summation.

Equation (2.7) is the most general form for the extra stress of the viscous shear flow. These were derived first by Reiner (1945) and then by Rivlin (1947). Therefore, such fluids are usually called as Reiner-Rivlin fluids. For incompressible materials I_1 equals zero and α, β and γ are considered as function of I_2 and I_3

It is recognised that (2.7) is far too general to solve a specific flow problem. In order to numerically solve different types of flow problems, there have been many constitutive models in the fluid flow literature proposed by investigators, such as, [1,2,3,4,5,6].

However, all models must include a special case in which the constitutive equation can be degenerated to

$$\boldsymbol{\tau} = \beta_0 \mathbf{d} \tag{2.9}$$

where β_0 is the constant. If we assume $\beta_0 = 2\mu_0$

$$\boldsymbol{\tau} = 2\mu_0 \mathbf{d} \tag{2.10}$$

where μ_0 is called *viscosity*. Materials obeying Equation (2.10) are the so called *Newtonian fluids.*

Substituting Equations (2.8) and (2.10) into Equation (2.2), we obtain,

$$\rho\frac{D\mathbf{v}}{Dt} = \rho\mathbf{g} + \mu_o \boldsymbol{\nabla}^2\mathbf{v} - \boldsymbol{\nabla}P, \tag{2.11}$$

which together with the continuity equation (2.5), are the 'Navier Stokes' equations valid for Newtonian fluid flow. For some fluids the magnitude of viscosity is independent of the rate of shear, and although it may vary considerably with temperature, it may be regarded as a constant for a particular fluid and temperature. Therefore, the Navier Stokes equations can be used in these limited cases.

There is, however, a fairly large category of liquids for which the velocity is not independent of the rate of shear and these liquids are referred to as non-Newtonian.

2.3.2   Inelastic generalised Newtonian flow

If we consider μ in Equation (2.10) as a function of the second invariant I_2 of **d** we can solve many more practical flow problems. Therefore we have,

$$\boldsymbol{\tau} = 2\mu(I_2)\mathbf{d} \qquad (2.12)$$

in which the Newtonian concept is retained and therefore it represents a generalised Newtonian fluid.

Considering equations (2.2) and (2.3) for generalised Newtonian flow, in the absence of body forces, the governing equations become

$$\rho\frac{\partial \mathbf{v}}{\partial t} = \nabla\!\cdot\!\mu\,\nabla\mathbf{v} - \rho\mathbf{v}\cdot\nabla\mathbf{v} - \nabla P \qquad (2.13)$$

and

$$\nabla\cdot\mathbf{v} = 0$$

2.3.3   Power law fluid

The most popular type of the generalised Newtonian flow is a power-law fluid. The constitutive equation of a power-law fluid may be written as

$$\boldsymbol{\tau} = 2\mu(I_2)\mathbf{d} = 2\mu_0 I_2^{\,p-1}\mathbf{d} \qquad (2.14)$$

where p is the power-law index, μ_0 is the consistency factor and $\mu(I_2)$ the shear viscosity.

There are two special cases of the equation (2.14): 1) $p{=}1$ and 2) $p{=}0$. If $p = 1$ then $\mu = \mu_0$, which is the Newtonian case. Considering the $p = 0$ case, Equation (2.14) becomes,

$$\mu(I_2) = \frac{\mu_0}{I_2} \qquad (2.15)$$

In fact I_2 is the effective deviatoric strain-rate $\dot{\bar{\epsilon}}$. If we re-define μ_0 as follows

$$\mu_0 = \frac{1}{3}\bar{\sigma} \qquad (2.16)$$

where $\bar{\sigma}$ is called effective stress, now we have,

$$\mu = \frac{1}{3}\frac{\bar{\sigma}}{\bar{\dot{\epsilon}}} \qquad (2.17)$$

One can immediately recognise that the above equation is the same as the one representing perfect plastic flow which obeys the Von-Mises yield criterion. On the other hand, if we consider the following hardening effect

$$\bar{\sigma} = Y(\bar{\dot{\epsilon}}) \qquad (2.18)$$

where Y is a monotonous increasing function which represents the material deformation behaviour. Apparently, from the above equation we can derive power-law constitutive relationship. For example, we can use

$$Y = \mu_0 \bar{\dot{\epsilon}}^{p-1}$$

and this viscoplastic function is suitable for simulating superplastic forming process.

Figures 2.1a to 2.1d show the development of the velocity profiles along a rectangular plane channel for $p=0$ (plastic flow) to $p=1$ (Newtonian flow). All velocities are non-dimensionalised by inlet values. Figure 2.2 shows that for the Newtonian case, an initially flat velocity profile becomes parabolic at the exit section, while for the power-law fluid, an initially parabolic velocity profile becomes flattened as p decreases.

Solutions often have a reduced viscosity when the rate of shear is large. That is so called shear thinning and such liquids are said to be *pseudo-plastic*. A few liquids exhibit the converse property of *dilatancy*, that is, their effective viscosity increases with increasing rate of shear.

In a viscous fluid a condition which must always be satisfied is that there should be no 'slipping' at solid boundaries. However, metals when close to their melting points can deform continuously under the action of a constant force, and thus in some degree behave like liquids of high viscosity. Their behaviour is non-Newtonian but with slipping at solid boundaries. Figure 2.3 shows the velocity profiles with different frictions at the slip boundaries (upper boundary of the channel).

2.4   Elastic Liquids

Another category of non-Newtonian fluids is elasto-viscous liquids or simply elastic liquids which possess a memory of past deformation. In a Newtonian fluid both the extra stress and the strain rate are constant therefore we can write,

$$\mathbf{d} = \frac{\boldsymbol{\tau}}{\eta} \tag{2.19}$$

where $\eta = 2\mu_0$. In elastic fluids where the extra stress is changed, a rapid increase of stress from $\boldsymbol{\tau}$ to $\boldsymbol{\tau} + \delta\boldsymbol{\tau}$ causes the material to be sheared through an additional angle $\frac{\delta\boldsymbol{\tau}}{G}$ where G represents an elastic modulus. The corresponding rate of shear is $\frac{1}{G} \cdot \frac{\partial \boldsymbol{\tau}}{\partial t}$ and so the total rate of shear in the material is

$$\mathbf{d} = \frac{\boldsymbol{\tau}}{\eta} + \frac{1}{G} \cdot \frac{\partial \boldsymbol{\tau}}{\partial t}$$

2.4.1   Maxwell model

The simple Maxwell model may given in the following form:

$$\boldsymbol{\tau} + \lambda_1 \frac{D}{Dt} \boldsymbol{\tau} = 2\mu_0 \mathbf{d} \tag{2.20}$$

where λ_1 is material constant and called *relaxation time*. μ_0 is constant viscosity, $\boldsymbol{\tau}$ is the extra-stress tensor, $\mathbf{d}$ the rate of deformation tensor.

In a convected coordinate system used to describe the governing equations, the stress and deformation tensors are functionals of the history of the metric tensor. Therefore, components of the stress and strain rate may be covariant (lower-convected) or contravariant (upper-convected). As examples in this text, upper-convected derivatives are considered. The constitutive equation can be expressed through a tensorially invariant relation between these tensors and their time derivatives. Therefore, the upper-convected Maxwell model can be written as follows

$$\boldsymbol{\tau} + \lambda_1 \overset{\triangledown}{\boldsymbol{\tau}} = 2\mu_0 \mathbf{D}$$

where ∇ denotes the upper-convected time derivative [2]. It can be seen that such a fluid has a constant viscosity.

2.4.2   Oldroyd's liquid B model

The standard viscoelastic model is derived by coupling springs and dashpots in series and parallel. The typical constitutive equation can be written as

$$\mathbf{T} + \lambda_1 \overset{\triangledown}{\mathbf{T}} = 2\mu_0(\mathbf{D} + \lambda_2 \overset{\triangledown}{\mathbf{D}})$$

in which $\mathbf{T}$, instead of τ is used for the extra-stress tensor, $\mathbf{d}$ is the deformation tensor and λ_2 is a constant retardation time. The parameter λ_2 is often regarded as expendable and the Oldroyd B model does not produce dramatically different results from the Maxwell model.

2.4.3   White-Metzner model

Both the Maxwell model and Oldroyd B model with a constant viscosity are far from satisfactory in their abilities to characterize flow problems dominated by shear viscosity. Considering both 'generalised Newtonian' model and the Maxwell model White and Metzner obtained the following equation

$$\mathbf{T} + \lambda \overset{\triangledown}{\mathbf{T}} = 2\mu(\mathbf{I_2})\mathbf{D}$$

where λ could be a function of I_2 and is called White-Metzner model [2].

2.5   Equation in Steady State Flow

On many occasions investigators are only interested in a fully developed flow regardless of the flow history. In this case the velocities do not change with time, although the fluid at each position may posses different velocities and therefore,

$$\frac{\partial}{\partial t} = 0 \tag{2.21}$$

As this derivative operates on parameters associated with flow. Therefore,

$$\frac{D}{Dt} = \frac{\partial}{\partial t} + \mathbf{v} \cdot \boldsymbol{\nabla} = \mathbf{v} \cdot \boldsymbol{\nabla}$$

which means that D/Dt is not equal to zero. $\mathbf{v} \cdot \boldsymbol{\nabla}$ is the advection term which is due to the change of velocity in the space. Steady flow can occur only if all the imposed conditions are constant in time. Therefore, the momentum equation for steady state flow has the following form:

$$\rho \mathbf{v} \cdot \boldsymbol{\nabla} \mathbf{v} = \rho \mathbf{g} + \boldsymbol{\nabla} \cdot \sigma \qquad (2.22)$$

with the same continuity equation as in (2.5).

For flows with low values of velocity the advective term

$$\rho \mathbf{v} \cdot \boldsymbol{\nabla} \mathbf{v} \qquad (2.23)$$

can be considered as an external force $\mathbf{F}$ calculated from velocities at the previous iteration [3].

2.6   Initial and Boundary Conditions

In order to solve the flow equations, one needs to determine the *initial conditions* at time $t = t_0$ in the domain Ω and the *boundary conditions* on the boundary S for any given problem.

Initial conditions:

The initial velocity field and pressure must be specified as

$$v_i(x_i, t = 0) = V_i^0(x_i) \qquad \text{in} \quad \Omega \qquad (2.24)$$

Boundary conditions:

The boundary conditions relevant to the class of problems to be tackled may be classified as below.

(a) Dirichlet or essential boundary conditions.

These are applicable to the fluid flow equations as specified velocities at the boundaries. These may be constant or be allowed to vary with time, *i.e.*

$$\mathbf{v} = f(x, y, t) \qquad \text{on } S_\mathbf{v} \qquad (2.25)$$

Pressure may not be specified at the boundaries as it is an implicit variable in an incompressible flow [7] which 'adjusts' itself to deliver a solenoidal velocity field. However, in the case of contained flow, *i.e.* specified velocities on all boundaries, the pressure becomes indeterminate and it must be specified at least at one point as a datum.

(b) Neumann or natural boundary conditions.

For the flow equations normal and tangential traction forces may be specified as below, say, on S_f

$$f_n = -P + 2\mu \frac{\partial v_n}{\partial n} \tag{2.26}$$

$$f_\tau = \mu \left(\frac{\partial v_n}{\partial \tau} + \frac{\partial v_\tau}{\partial n} \right) \tag{2.27}$$

where n and τ are the unit normal and tangent vectors with respect to the boundary S_f.

S_v and S_f are parts of the boundary S of the computational domain Ω, in a way that the following relations hold,

$$S_v \cup S_f = S$$

$$S_v \cap S_f = \phi$$

where ϕ is the null set.

2.7   Stream Function

For two-dimensional or axisymmetric laminar flow, the fluid particles follow a series of lines called streamlines and their velocities are tangential to them. Considering only time independent two-dimensional incompressible flow, the continuity equation is (as before)

$$\frac{\partial u}{\partial x} + \frac{\partial v}{\partial y} = 0 \tag{2.28}$$

This can be satisfied automatically by choosing

$$u_x = -\frac{\partial \Psi}{\partial y}$$

$$u_y = \frac{\partial \Psi}{\partial x} \qquad (2.29)$$

where $\Psi = \Psi(x, y, t)$ is a function of the indicated variables.

The quantity Ψ is known as the *stream function*. The curves $\Psi = const.$ are the streamlines which are always parallel to the velocity ($\mathbf{u}$). For fixed boundaries in two dimensional flow the boundary itself is a streamline. Therefore along a fixed boundary, we have

$$\Psi = const.$$

The stream function plays an important role in two dimensional flows. As no flow occurs across streamlines instantaneously therefore the flux of liquid per unit length in z direction equals an increase in value of the stream function between two streamlines.

In order to obtain the stream function we substitute (2.29) into continuity equation (2.28) and have

$$\frac{\partial^2 \Psi}{\partial x^2} + \frac{\partial^2 \Psi}{\partial y^2} = 0 \qquad (2.30)$$

which is called as Laplace equation.

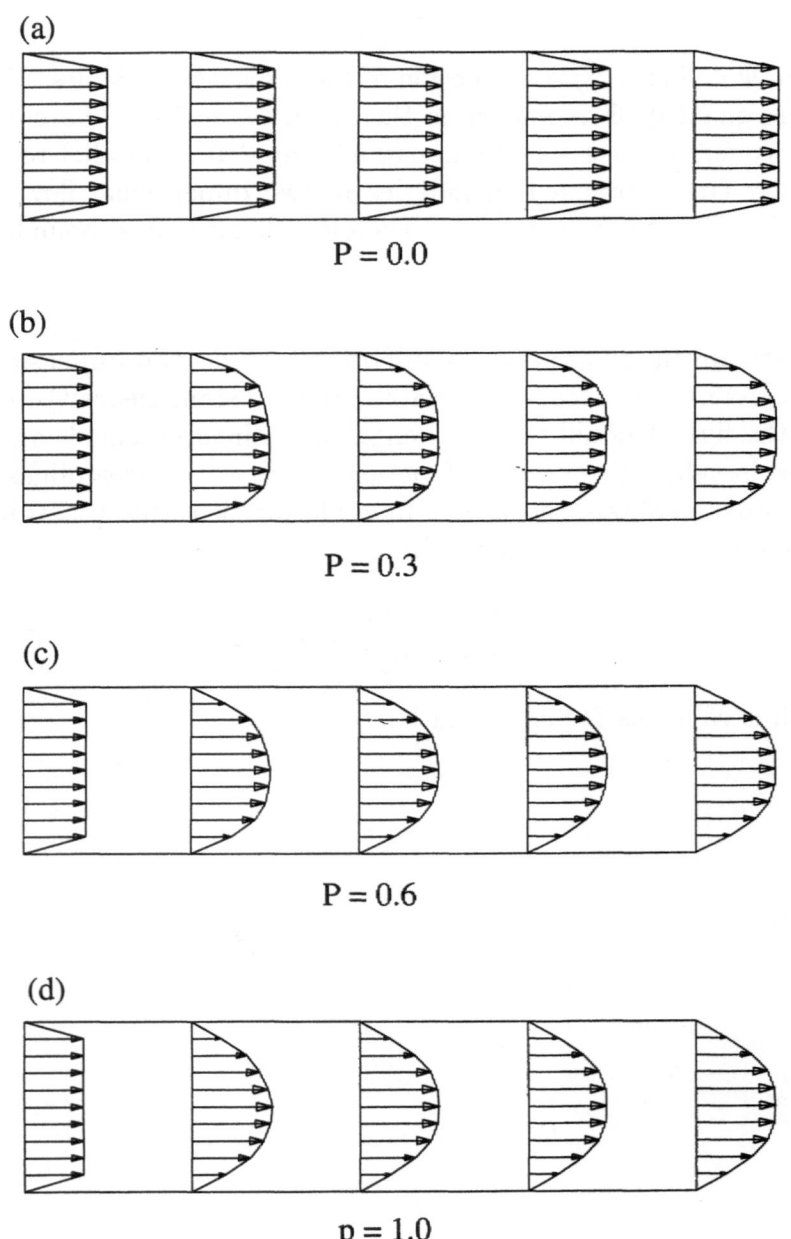

Figure 2.1: The development of the velocity profile along a rectangular plane channel with variation of power law index.

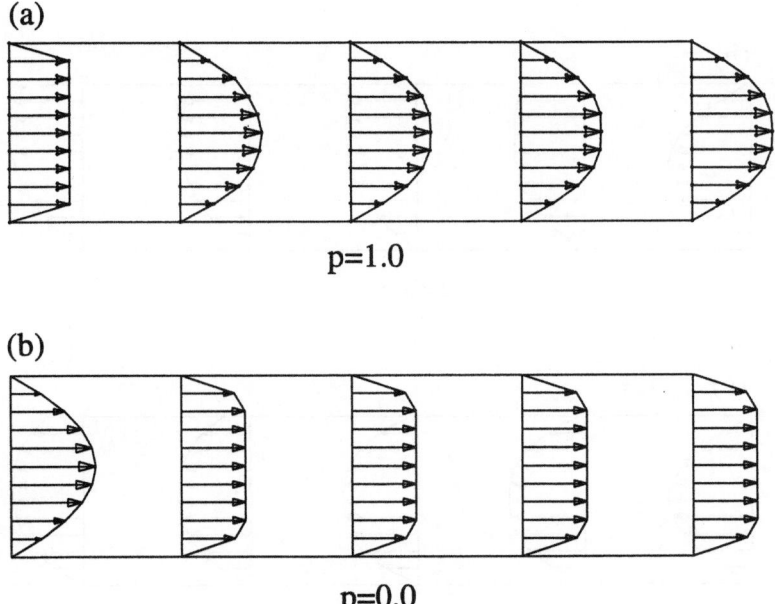

Figure 2.2: Two typical cases, (a) Newtonian flow with p=1; (b) power-law fluid with p=0

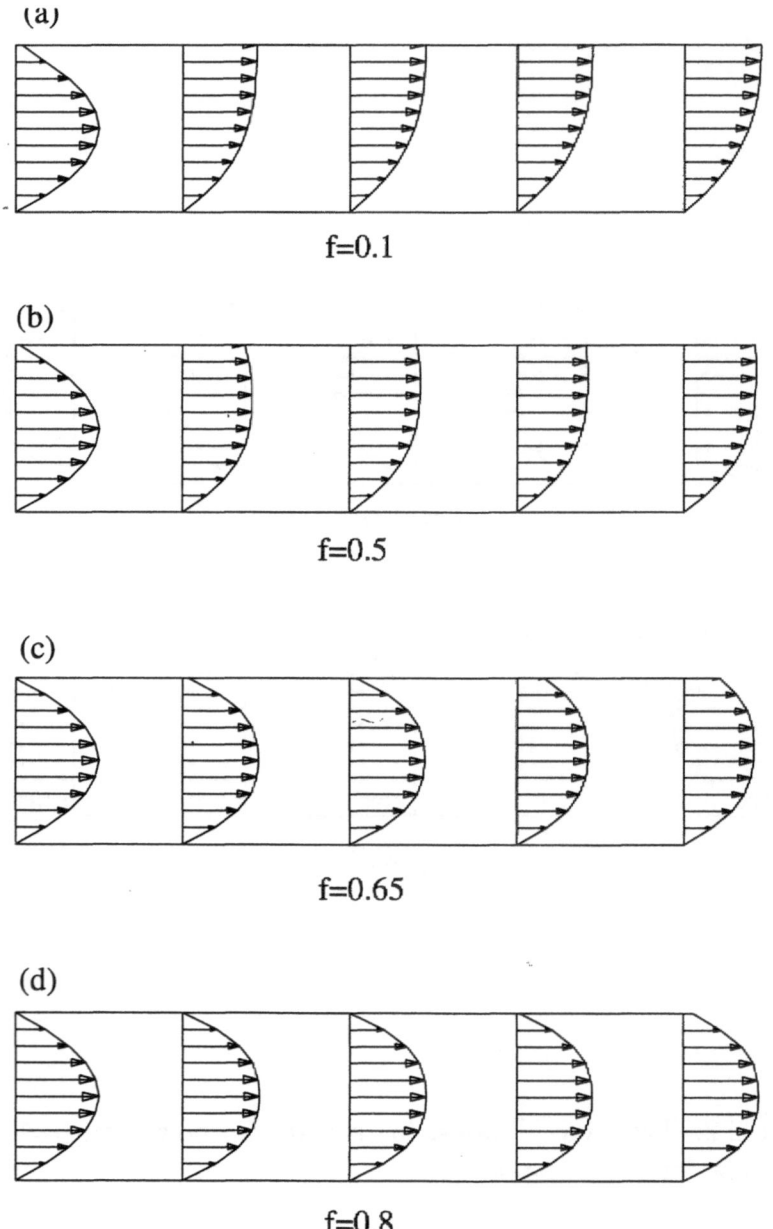

Figure 2.3: The development of the velocity profile along a rectangular plane channel with different frictions

Bibliography

[1] S.C.Hunter. *Mechanics of Continuous Media*. Ellis Horwood Ltd, Sussex, U.K., 1976.

[2] M.J.Crochet. *Numerical Simulation of Non-Newtonian Flow*. Elsevier, New York, U.K., 1984.

[3] C.Taylor and T.G.Hughes. *Finite Element Programming of the Navier-Stokes Equations*. Pineridge Press, Swansea, U.K., 1981.

[4] J.J. Connor and C.A.Brebbia. *Finite Element Techniques for Fluid Flow*. Newnes-Butterworths, London, U.K., 1976.

[5] T.J.Chung. *Finite Element Analysis in Fluid Dynamics*. McGRAW-HILL, New York, U.S.A., 1978.

[6] R.Peyret and T.D.Taylor. *Computational Methods for Fluid Flow*. Springer-Verlag, New York, USA, 1983.

[7] P.M.Gresho, R.L.Lee, and R.L.Sani. On the time-dependent solution of the incompressible Navier-Stokes equations in two and three dimensions. In *Recent Advances in Numerical Methods in Fluids*, Pineridge Press Limited, Swansea, 1980.

Chapter 3

Finite Element Method

3.1 Introduction

If the fluid flow domain and boundary conditions are well posed then the Navier-Stokes equations can be analytically solved, however this is possible only for the simplest type of problems.

One simple but instructive problem is that flow between infinite parallel plates (see Figure 3.1). The bottom plate is fixed, while the upper plate is moving at speed of U in x direction.

The flow is assumed to be fully developed and therefore the steady-state equations are applied with the following boundary conditions:

$$
\begin{aligned}
u(x, z) &= 0 & \text{at} \quad z &= 0 \\
u(x, z) &= U & \text{at} \quad z &= h \\
v(x, z) &= w(x, z) = 0 & \text{for all } x \text{ and } 0 \le z \le h
\end{aligned}
$$

For this special case the Navier-Stokes equations (2.4) can be written as,

$$
\frac{\partial u}{\partial x} = 0 \tag{3.1}
$$

$$
\frac{-1}{\rho}\frac{\partial P}{\partial x} + \mu\frac{\partial^2 u}{\partial z^2} = 0
$$

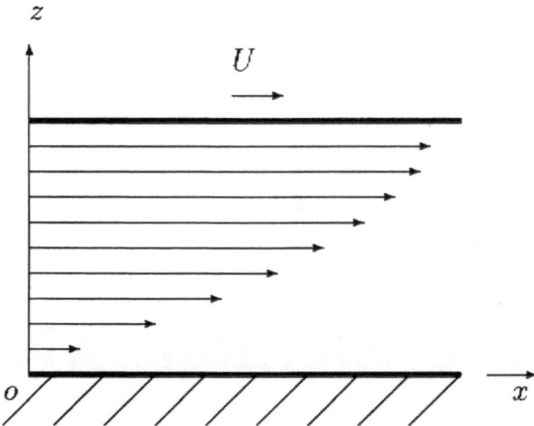

Figure 3.1: Flow between infinite parallel plates

$$-g - \frac{1}{\rho}\frac{\partial P}{\partial z} = 0$$

Integrating the third equation of (3.1), we have

$$P = -\rho g z + f(x, z)$$

It is noted that P is linear along z direction and $\frac{\partial P}{\partial x}$ is independent of z. From $\frac{\partial u}{\partial x} = 0$, it is known that u is only a function of z. Integrating the second equation of (3.1) twice, we obtain

$$u = \frac{1}{2\mu}\frac{\partial P}{\partial x}z^2 + Az + B$$

From the boundary conditions $u = 0$ at $z = 0$ and $u = U$ at $z = h$, we have

$$B = 0$$
$$A = \frac{U}{h} - \frac{1}{2\mu}\frac{\partial P}{\partial x}h,$$

therefore, we obtain velocity u as

$$u = \frac{Uz}{h} - \frac{1}{2\mu}\frac{\partial P}{\partial x}z(h - z)$$

If the upper plate is also fixed, that is $U = 0$, then we have

$$u = -\frac{1}{2\mu}\frac{\partial P}{\partial x}z(h - z)$$

and

$$u_{max} = -\frac{h^2}{8\mu}\frac{\partial P}{\partial x}$$

where u_{max} is the maximum velocity (*i.e.* the centre-line velocity).

This above represents one of the simplest examples of fluid flow problems. There exist no general methods to solve the Navier-Stokes equations theoretically even for Newtonian flow problems, which is only a particular case of viscous fluid flow.

The Navier-Stokes equations do however provide very useful information which can lead to a considerable reduction in the quantity of experimental work required for studying flow behaviour in any given problem. Furthermore, the Navier-Stokes equations provide a great basis for developing numerical and computational procedures which then allow accurate modelling a large variety of engineering flow problems.

It is possible to model (with varying degrees of accuracy) almost any practical fluid flow problem using the phenomenal computational resources available to modern engineers and scientists. Computer programs for solving fluid flow problems require the governing differential equations to be recast in an algebraic form, which are than solved by the computer involving only the basic arithmetic operations. To achieve this, various forms of *discretisation* of the continuum problem defined by the differential equations can be used. In the continuum problem the solution is satisfied at all points in the problem domain. The discretised form of the problem only requires the solution to be satisfied at a finite number of points in the domain. In the remainder of the domain appropriate interpolation functions are used. Of the various forms of discretisation which are possible, one of the simplest is the *finite difference method* which has been used for many decades. Today, the *finite element method* developed in the 1960's is increasingly being seen as the most attractive tool for approximate solutions of all kinds of equations. Finite element formulations can be derived from several different approaches, such as variational principles and weighted residuals.

3.2 Variational Principle and Rayleigh-Ritz Method

Some physical problems can be stated directly in the form of a variational principle which consists of determining the function which makes a certain integral statement, called *functional*, stationary. However, the form of the variational principle is not always obvious and, indeed, such a principle does not exist for many continuum problems for which well-defined differential equations may be formulated. In the followings text we will examine the variational principle for the Poisson equation. Consider the situation of two space dimensions (x, y) and the functional defined by [1]

$$\Pi(T) = \int_\Omega \frac{1}{2} \left[\left(\frac{\partial u}{\partial x}\right)^2 + \left(\frac{\partial u}{\partial y}\right)^2 + f(x,y)u \right] d\Omega \qquad (3.2)$$

where $f(x, y)$ is an unknown function.

It can be shown that Equation (3.2) is equivalent to the following differential equation

$$\frac{\partial}{\partial x}\left(\frac{\partial u}{\partial x}\right) + \frac{\partial}{\partial y}\left(\frac{\partial u}{\partial y}\right) - f(x,y) = 0 \qquad (3.3)$$

or

$$\frac{\partial^2 u}{\partial x^2} + \frac{\partial^2 u}{\partial y^2} = f(x,y) \qquad (3.4)$$

or using the operator *del* (∇), we can write

$$\nabla^2 u = f(x,y) \qquad (3.5)$$

The above expression is known as the Poisson equation and integral in Equation (3.2) is called the *variational principle* for the Poisson equation.

Applying the variational operation to (3.2), we have

$$\delta\Pi(u) = \int_\Omega \left[\frac{1}{2}\left(\frac{\partial u}{\partial x}\right) \frac{\partial}{\partial x}(\delta u) + \frac{1}{2}\left(\frac{\partial u}{\partial y}\right) \frac{\partial}{\partial y}(\delta u) - f(x,y)\delta u \right] d\Omega$$
$$= 0 \qquad (3.6)$$

We may rewrite Equation (3.6), using Green's theorem, as follows

$$\int_\Omega A\frac{\partial B}{\partial x} dx dy = -\int_\Omega \frac{\partial A}{\partial x} B dx dy + \oint_S AB n_x dS$$

$$\int_\Omega A\frac{\partial B}{\partial y} dx dy = -\int_\Omega \frac{\partial A}{\partial y} B dx dy + \oint_S AB n_y dS \qquad (3.7)$$

where A and B are suitable differential functions, n_x and n_y are the direction cosines of the outward normal n to the closed curve S surrounding an area Ω in the (x,y) plane. The integration around S is made in an anticlockwise direction and it should noted that

$$n_x\frac{\partial A}{\partial x} + n_y\frac{\partial A}{\partial y} = \frac{\partial A}{\partial n} \qquad (3.8)$$

Using Equations (3.7) and (3.8) in Equation (3.6), we obtain

$$\delta\Pi(u) = \int_S \frac{\partial u}{\partial n}\delta u dS - \int_\Omega \left[\frac{\partial^2 u}{\partial x^2} + \frac{\partial^2 u}{\partial y^2} + f(x,y) \right] \delta u d\Omega = 0 \quad (3.9)$$

where δu is arbitrary. It is necessary when Π is stationary that

$$\frac{\partial u}{\partial n} = 0 \qquad (3.10)$$

and

$$\frac{\partial^2 u}{\partial x^2} + \frac{\partial^2 u}{\partial y^2} + f(x,y) = 0 \qquad (3.11)$$

The differential Equation (3.11) is called the *Euler-Lagrange equation* and Equation (3.10) as the *natural* boundary condition.

It can be seen that instead of solving the differential equation (Euler-Lagrange equation) directly, we can minimize its functional to approach the problem of the two-dimensional Poisson equation with the natural boundary conditions on S.

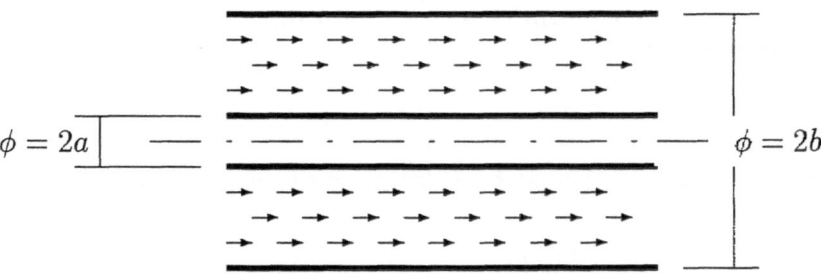

Figure 3.2: Axisymmetric shear flow along an annulus

Let us now consider a specific problem, which is an axisymmetric flow along the annulus of two parallel cylindrical walls with inner radius $r = a$ and outer radius $r = b$ (see Figure 3.2). In this example, the pressure gradient, dP_z/dz, in the z direction is a constant ($-m$, say) and the velocity, w, in the z direction depends only on the radial coordinate r. The governing equation for this problem is an ordinary differential equation,

$$\frac{1}{r}\frac{\partial}{\partial r}\left(r\frac{\partial w}{\partial r}\right) = -m/\mu \qquad (3.12)$$

with the boundary conditions of

$$w(r) = 0 \qquad \text{at} \quad r = a$$

and

$$w(r) = 0 \qquad \text{at} \quad r = b \qquad (3.13)$$

The classical analytical solution to the flow problem is presented in (**) which has the form

$$w(r) = \frac{m}{4\mu}\left[(b^2 - r^2) - (b^2 - a^2)\frac{ln(\frac{b}{r})}{ln(\frac{b}{a})}\right] \qquad (3.14)$$

for the rectilinear shear flow under pressure gradient m in the annulus $b \geq r \geq a$. The functional to this problem can be written as follows

$$\Pi(w) = \frac{1}{2} \int_0^{2\pi} \int_a^b \left[\left(\frac{\partial w}{\partial r} \right)^2 - \frac{2m}{\mu} w \right] r \, dr \, d\theta \qquad (3.15)$$

We now consider the Rayleigh-Ritz method which uses *trial functions* $N(r)_n$ to approximate the real solution $w(r)$ in the functional. That is

$$\bar{w}(r) = C_0 + \sum_{n=1}^{\infty} C_n N(r)_n, \qquad (3.16)$$

which must satisfy the essential boundary conditions. Then the functional has to be minimized with respect to the parameters C_n. In the present problem, for instance, we can use a power series expansion of the form

$$\bar{w}(r) = C_0 + \sum_{n=1}^{\infty} C_n r^n. \qquad (3.17)$$

For simplicity, we consider the case where $n = 2$, thus

$$\bar{w}(r) = C_0 + C_1 r + C_2 r^2 \qquad (3.18)$$

Note that boundary condition $w(b) = 0$ which gives $C_0 = -C_1 b - C_2 b^2$ and

$$\bar{w}(r) = C_1(r - b) + C_2(r^2 - b^2) \qquad (3.19)$$

Further consideration of $w(a) = 0$, giving $C_1 = -C_2(b + a)$, we have

$$\bar{w}(r) = C_2(r - a)(r - b) \qquad (3.20)$$

and

$$\frac{d\bar{w}(r)}{dr} = C_2(2r - a - b). \qquad (3.21)$$

By substituting Equations (3.20) and (3.21) into the functional (3.15) we obtain

$$\Pi(w) = \pi \int_a^b \left[\left(\frac{d\bar{w}}{dr} \right)^2 - \frac{2m}{\mu} \bar{w} \right] r \, dr \qquad (3.22)$$

then

$$\Pi(w) = \pi C_2^2 \left[(b^4 - a^4) - \frac{4}{3}(b + a)(b^3 - a^3) + \frac{1}{2}(b + a)^2(b^2 - a^2) \right] -$$

$$\frac{2\pi m C_2}{\mu}\left[\frac{1}{4}(b^4 - a^4) - \frac{1}{3}(b + a)(b^3 - a^3) + \frac{1}{2}ab(b^2 - a^2)\right] \quad (3.23)$$

For Π to be a minimum, that is

$$\frac{\partial \Pi}{\partial C_n} = 0, \qquad n = 2 \qquad (3.24)$$

which provides

$$C_2\left[(b^4 - a^4) - \frac{4}{3}(b + a)(b^3 - a^3) + \frac{1}{2}(b + a)^2(b^2 - a^2)\right] =$$

$$\frac{m}{\mu}\left[\frac{1}{4}(b^4 - a^4) - \frac{1}{3}(b + a)(b^3 - a^3) + \frac{1}{2}ab(b^2 - a^2)\right] \qquad (3.25)$$

If $a = 1$ and $b = 2$, we have

$$C_2 = -\frac{m}{2\mu} \qquad (3.26)$$

Substituting Equation (3.26) into Equation (3.20) yields

$$\bar{w}(r) = \frac{-m}{2\mu}(r - a)(r - b) \qquad (3.27)$$

From (3.21) and assuming that

$$\frac{d\bar{w}(r)}{dr} = 0, \qquad (3.28)$$

then we obtain the maximum value of w when $r = 0.5(a + b) = 1.5$

$$\bar{w}(1.5) = 0.125\frac{m}{\mu} \qquad (3.29)$$

whereas Equation (3.14) gives the analytical solution of $w(1.5) = 0.1262\frac{m}{\mu}$. As we increase n we expect to converge to this solution [2].

3.3 Galerkin Weighted Residual Method

As mentioned in the previous section, for many continuum problems, a suitable variational principle is not available, since no corresponding functionals exist although their differential equations may well be formulated. As an alternative to solve such differential equations, we may use a variety of weighted residual methods. Let us consider the two dimensional problem of (3.5). We begin by introducing the error, or *residual*, R_Ω in the approximation which is defined by

$$R_\Omega = \frac{\partial}{\partial x}(\frac{\partial \bar{u}}{\partial x}) + \frac{\partial}{\partial y}(\frac{\partial \bar{u}}{\partial y}) - f(x, y), \qquad (3.30)$$

where $\bar{u}$ contains trial functions and satisfies the Dirichlet boundary condition of $\bar{u} = u_0$ at S_u. The smaller the residual, the better the approximation. It should be noted that R_Ω is a function of position in Ω. Now, we attempt to reduce this residual as close to zero as possible. If we have

$$\int_\Omega W_i R_\Omega d\Omega = 0, \qquad i = 1, 2, ..., M \qquad (3.31)$$

where W_i is a set of arbitrary functions and $M \to \infty$, then it can be said that the R_Ω vanishes. We can adjust the free parameters C_i in $\bar{u}$ to approach this objective. Here W_i are called *weighting functions*. Expanding Equation (3.31), we have

$$\int_\Omega W_i \left[\frac{\partial}{\partial x}\left(\frac{\partial \bar{u}}{\partial x}\right) + \frac{\partial}{\partial y}\left(\frac{\partial \bar{u}}{\partial y}\right) - f(x, y) \right] dx dy = 0 \qquad (3.32)$$

A function $\bar{u}(x, y)$ that satisfies Equation (3.32) for every function W_i in Ω is a weak solution of the differential equation, whereas the strong solution $u(x, y)$ satisfies the differential equation at every point of Ω.

Equation (3.32) may be rewritten, using Green's theorem, to give

$$-\int_\Omega \left[\frac{\partial W_i}{\partial x}\frac{\partial \bar{u}}{\partial x} + \frac{\partial W_i}{\partial y}\frac{\partial \bar{u}}{\partial y} \right] dx dy + \int_\Omega W_i f(x, y) dx dy +$$

$$\int_S \frac{\partial \bar{u}}{\partial n} W_i dS = 0 \tag{3.33}$$

Since both W_i are arbitrary, we can limit the choice of the weighting functions as

$$W_i = 0 \quad \text{on} \quad S \tag{3.34}$$

Now, it can be seen that the term involving the weighted integral of $\frac{\partial \bar{u}}{\partial n}$ on the boundary vanishes and the approximating equation becomes

$$\int_\Omega \left[\frac{\partial W_i}{\partial x} \frac{\partial \bar{u}}{\partial x} + \frac{\partial W_i}{\partial y} \frac{\partial \bar{u}}{\partial y} \right] dx\, dy - \int_\Omega W_i f(x, y) dx\, dy = 0 \tag{3.35}$$

This is called as *weak form* of the Poisson equation. For the variational method, the boundary condition

$$-\frac{\partial u}{\partial n} = 0 \quad \text{on} \quad S \tag{3.36}$$

is in some way a natural boundary condition for this problem as the formulation eliminates the need for an actual evaluation of $\frac{\partial \bar{u}}{\partial n}$ on the boundaries and such boundaries do not enter explicitly into Equation (3.35).

In order to utilize the weak form Equation (3.35) to obtain the approximation of the solution, one has to first choose appropriate trial functions, $N_i(x, y)$ to represent the real solution, that is

$$\bar{u}(x, y) = C_0 + \sum_{i=1}^{\infty} C_i N_i(x, y) \tag{3.37}$$

which must satisfy the essential boundary conditions. The second task is to choose the weighting functions. The most popular weighted residual method is where the trial functions themselves are chosen as the weighting functions, thus

$$W_i(x, y) = N_i(x, y) \tag{3.38}$$

This method was first used by Galerkin and is referred to the Galerkin method. Therefore, for Galerkin weighted residual method, the weak form of the 2-D Poisson equation may be written as

$$\int_\Omega \left[\frac{\partial N_i}{\partial x} \frac{\partial \bar{u}}{\partial x} + \frac{\partial N_i}{\partial y} \frac{\partial \bar{u}}{\partial y} \right] dx\, dy - \int_\Omega N_i f(x, y) dx\, dy = 0 \tag{3.39}$$

Let us consider the same problem as given by Equations (3.12) and (3.13) in Section 3.2. The residual in Ω is

$$R_\Omega = \frac{1}{r}\frac{\partial}{\partial r}\left(r\frac{\partial \bar{u}}{\partial r}\right) \tag{3.40}$$

The weak form in this case can be written as

$$\int_0^{2\pi}\int_a^b\left[\frac{\partial W_i}{\partial r}\frac{\partial \bar{u}}{\partial r}\right]r\,dr\,d\theta = 0. \tag{3.41}$$

In order to obtain the solution, we use the two term approximation as in Equation (3.20) for $\bar{u}$, that is

$$\bar{u}(r) = C_2(r - a)(r - b) \tag{3.42}$$

which satisfies the essential boundary condition. According to the Galerkin method, the weighting function is

$$W_2 = (r - a)(r - b) \tag{3.43}$$

which vanish on the S, that is $r = a$ and $r = b$. Substitution of Equations (3.42) and (3.43) into the weak form Equation (3.41) yields

$$C_2\left[(b^4 - a^4) - \frac{4}{3}(b + a)(b^3 - a^3) + \frac{1}{2}(b + a)^2(b^2 - a^2)\right] =$$

$$\frac{m}{\mu}\left[\frac{1}{4}(b^4 - a^4) - \frac{1}{3}(b + a)(b^3 - a^3) + \frac{1}{2}ab(b^2 - a^2)\right] \tag{3.44}$$

which is identical to Equation (3.25). Using the same procedure the approximate solution can be obtained. Thus we see here that the Rayleigh-Ritz and Galerkin methods lead to identical results. For generality, we will use the Galerkin weighted residual approach to introduce the finite element method.

3.4 Virtual Work Principle

For the steady state flow, the governing equation can be described as follows

$$\nabla \cdot \sigma + \mathbf{b} = 0 \tag{3.45}$$

in which σ is the vector of stresses and **b** is the vector of body forces. **b** may be expressed as $\rho g - \rho \mathbf{v} \cdot \nabla \mathbf{v}$ if the velocity is considerably low. The relationship between stress and rate of strain is given by Stokes' law of friction in fluids as

$$\sigma_{ij} = -P\delta_{ij} + \tau_{ij}$$
$$\tau_{ij} = 2\mu(I_2)d_{ij} \qquad (3.46)$$

where τ_{ij} is the extra stress, d_{ij} is the rate of strain and I_2 is the second invariant of the strain rate. The virtual work principle can be written as

$$\int_\Omega \delta \mathbf{d}^{\mathbf{T}} \sigma d\Omega + \int_\Omega \delta \mathbf{v}^{\mathbf{T}} \mathbf{b} d\Omega = \int_{S_t} \delta \mathbf{v}^{\mathbf{T}} \mathbf{t} dS_t \qquad (3.47)$$

where $\delta \mathbf{v}$ is the vector of virtual velocities, $\delta \mathbf{d}$ is the vector of associated virtual strain rates, S_t is that part of the boundary on which boundary tractions are prescribed and S_v is that part of the boundary on which velocities are prescribed, that is $\delta \mathbf{v} = \mathbf{0}$ at S_v. The virtual work principle (3.47) is equivalent to the steady state equation of (3.45). Therefore, from the virtual work principle one can obtain a numerical solution for (3.45) which is shown in the following chapter. In order to impose the continuity equation, an additional term has to be considered. It has been shown that a penalty function statement can be constructed by defining the pressure as follows:

$$\lambda \nabla \cdot \mathbf{v} = -\mathbf{P} \qquad (3.48)$$

in which λ is a penalty parameter. If λ is sufficiently large, the incompressibility condition can be satisfied. It is noted that for a compressible plastic flow, λ is the volumetric modulus.

The penalty function statement (3.48) may be written in the virtual work form as

$$\int_\Omega \delta \mathbf{d}_{\mathbf{v}}^{\mathbf{T}} (\lambda \nabla \cdot \mathbf{v} + \mathbf{P}) d\Omega = \mathbf{0} \qquad (3.49)$$

where

$$\delta \mathbf{d_v} = \delta \nabla \cdot \mathbf{v} = \nabla \cdot \delta \mathbf{v} \qquad (3.50)$$

Then Equation (3.49) becomes

$$\int_\Omega [\nabla \cdot \delta \mathbf{v}]^T (\lambda \nabla \cdot \mathbf{v} + \mathbf{P}) d\Omega = \mathbf{0} \qquad (3.51)$$

Now, the left-hand side of Equation (3.51) is multiplied by an arbitrary constant α and added to Equation (3.47) to obtain the penalty function statement. Since α is an arbitrary constant, say $\alpha = 1$, the total virtual work statement becomes

$$\int_{\Omega} [\delta \mathbf{d}]^T \sigma \mathrm{d}\Omega \ + \ \int_{\Omega} [\delta \mathbf{v}]^T \mathbf{b} \mathrm{d}\Omega - \int_{S_t} [\delta \mathbf{v}]^T \mathbf{t} \mathrm{d} S_t +$$
$$\int_{\Omega} [\boldsymbol{\nabla} \cdot \delta \mathbf{v}]^T (\lambda \boldsymbol{\nabla} \cdot \mathbf{v} + \mathbf{P}) \mathrm{d}\Omega = \mathbf{0} \quad (3.52)$$

3.5 Finite Element Method in Two Dimensions

3.5.1 Introduction

In the previous sections we employed a single expression valid throughout the whole domain Ω to approximate the real solution of Poisson equation, that is

$$\bar{u}(x,y) = C_0 + \sum_{i=1}^{\infty} C_i N_i(x,y) \quad (3.53)$$

The integrals of the approximating equations, such as Equation (3.2) and Equation (3.35), were evaluated in one operation over this domain. An alternative approach is to divide the domain Ω into a number of nonoverlapping *subregions* or *finite elements*, Ω_e. The shape of the finite elements is generally restricted to simple polygons such as triangles and quadrilaterals in two dimensions, pyramids and triangular and rectangular prisms in three dimensions, and so on. The approximation $\bar{u}$ is constructed in a *piecewise* manner over such elements. In each element, the approximation solution assumes the form of a linear combination of prescribed functions, thus

$$\begin{aligned}
\bar{u}(x,y) \ &= \ \sum_{\Omega_1}^{i=1,J} C_i^{(1)} N_i^{(1)}(x,y) + \sum_{\Omega_2}^{i=1,J} C_i^{(2)} N_i^{(2)}(x,y) + ... \\
&+ \ \sum_{\Omega_M}^{i=1,J} C_i^{(M)} N_i^{(M)}(x,y) \quad (3.54)
\end{aligned}$$

where J is the number of free parameters C_i^e in each element and M is the total number of elements. In such a case, the definite integrals occurring in the weak form Equation (3.39) can be obtained simply by summing the contributions from each element as

$$\sum_{e=1}^{M} \int_{\Omega_e} \left[\frac{\partial N_i^e}{\partial x} \frac{\partial \bar{u}^e}{\partial x} + \frac{\partial N_i^e}{\partial y} \frac{\partial \bar{u}^e}{\partial y} \right] dx dy - \sum_{e=1}^{M} \int_{\Omega_e} N_i^e f(x,y) dx dy = 0$$

(3.55)

where

$$\bar{u}^e(x,y) = \sum_{\Omega_e}^{i=1,J} C_i^e N_i^e(x,y)$$

(3.56)

and Ω_e is the area of an element.

The shapes of elements chosen are usually the same for a particular problem and the definition of trial functions over each element can be made in a repeatable manner. Therefore, we can work out the formulation in one element Ω_e and quite readily extend it to the whole domain Ω. It is here that the essential idea of the finite element method lies. In fact the example used in Sections 3.2 and 3.3 is a special case of the finite element method where a single element is used.

3.5.2   K Matrix and load vector

For each element, we have to evaluate the following integral

$$\int_{\Omega_e} \left[\frac{\partial N_i^e}{\partial x} \frac{\partial \bar{u}^e}{\partial x} + \frac{\partial N_i^e}{\partial y} \frac{\partial \bar{u}^e}{\partial y} \right] dx dy - \int_{\Omega_e} N_i^e f(x,y) dx dy$$

$$(i = 1, J)$$

(3.57)

Now, we assume that the $N_i^e(x,y)$ are analytically simple, such as low-order polynomials and the like and C_i^e are local or nodal values of the solution u, that is

$$\bar{u}^e(x,y) = \sum_{\Omega_e}^{j=1,J} u_j^e N_j^e(x,y)$$

(3.58)

Substituting Equation (3.58) into Equation (3.57), we obtain

$$\int_{\Omega_e} \left[\frac{\partial N_i^e}{\partial x} \frac{\partial N_j^e}{\partial x} + \frac{\partial N_i^e}{\partial y} \frac{\partial N_j^e}{\partial y} \right] u_j^e dx dy - \int_{\Omega_e} N_i^e f(x,y) dx dy$$

$$(i, j = 1, J) \qquad (3.59)$$

Equation (3.59) may be rewritten as

$$K_{ij}^e u_j^e - f_i^e \qquad (i, j = 1, J) \qquad (3.60)$$

where

$$K_{ij}^e = \int_{\Omega_e} \left[\frac{\partial N_i^e}{\partial x} \frac{\partial N_j^e}{\partial x} + \frac{\partial N_i^e}{\partial y} \frac{\partial N_j^e}{\partial y} \right] dx dy \qquad (3.61)$$

which is named as the element **K** *matrix* and

$$f_i^e = \int_{\Omega_e} N_i^e f(x, y) dx dy \qquad (i = 1, J) \qquad (3.62)$$

which is called the element *load vector*. Therefore, the approximating equations can be written as

$$K_{ij} u_j = f_i \qquad (i, j = 1, Q) \qquad (3.63)$$

where

$$K_{ij} = \sum_{e=1}^{M} K_{ij}^e \qquad (3.64)$$

$$f_i = \sum_{e=1}^{M} f_i^e \qquad (3.65)$$

and Q is the total number of unknowns, u_j. So, the global matrix K_{ij} and load vector f_i are assembled from each element.

In order to discuss further details, the type of finite element to be used for analysis must be defined.

3.5.3 Triangular elements

The simplest finite element geometry employed in two dimensions is that obtained by covering Ω with triangles having nodes (or solution points) at the vertices (Figure 3.3) and in which the finite element solution $\bar{u}(x, y)$ consists of plane triangular 'tiles'. The triangular finite element in two spatial dimensions holds an important place, particularly for the purpose of adaptively remeshing, because it is very flexible in idealising an arbitrary domain Ω geometrically.

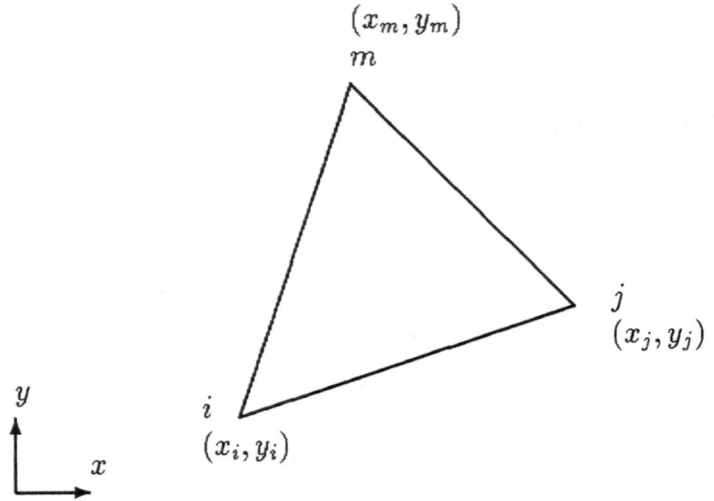

Figure 3.3: 3 node triangular element

Here, the linear triangular element is used to demonstrate all essential features of two dimensional finite element analysis. The element has three nodes numbered anticlockwise as 1,2,3, or generally as i, j, m. A linear variation of solution u^e is assumed throughout the element, each node has one degree of freedom corresponding to the free parameters, that is the unknown value u_i^e at that point. Therefore the solution at any coordinate position x, y within the element can be represented as

$$u^e = a_1 + a_2 x + a_3 y \qquad (3.66)$$

since

$$u^e(x_1, y_1) = u_1^e$$

$$u^e(x_2, y_2) = u_2^e$$

$$u^e(x_3, y_3) = u_3^e \qquad (3.67)$$

so, we have

$$u_1^e = a_1 + a_2 x_1 + a_3 y_1$$

$$u_2^e = a_1 + a_2 x_2 + a_3 y_2$$

$$u_3^e = a_1 + a_2 x_3 + a_3 y_3 \qquad (3.68)$$

Solving Equation (3.68) for a_1, a_2 and a_3 yields

$$a_1 = \frac{1}{2A}[(x_2 y_3 - x_3 y_2)u_1^e + (x_3 y_1 - x_1 y_3)u_2^e + (x_1 y_2 - x_2 y_1)u_3^e]$$

$$a_2 = \frac{1}{2A}[(y_2 - y_3)u_1^e + (y_3 - y_1)u_2^e + (y_1 - y_2)u_3^e]$$

$$a_3 = \frac{1}{2A}[(x_3 - x_2)u_1^e + (x_1 - x_3)u_2^e + (x_2 - x_1)u_3^e] \qquad (3.69)$$

where

$$A = \frac{1}{2}det\begin{bmatrix} 1 & x_1 & y_1 \\ 1 & x_2 & y_2 \\ 1 & x_3 & y_3 \end{bmatrix} = \qquad \text{element area} \qquad (3.70)$$

Therefore, we write

$$u^e = N_1^e u_1^e + N_2^e u_2^e + N_3^e u_3^e \qquad (3.71)$$

where

$$N_1^e = \frac{1}{2A}(b_1 + c_1 x + d_1 y)$$

$$N_2^e = \frac{1}{2A}(b_2 + c_2 x + d_2 y)$$

$$N_3^e = \frac{1}{2A}(b_3 + c_3 x + d_3 y) \qquad (3.72)$$

in which

$$b_i = x_j y_m - x_m y_j$$

$$c_i = y_j - y_m$$

$$d_i = x_m - x_j \qquad (3.73)$$

and i, j and m are taken as $1, 2, 3$ in cyclic permutation. Here, the element trial functions N_i^e are named as the *shape functions*. It should be noted that $N_i^e(x, y)$ has a unit value at (x_i, y_i) of node i, whereas it has zero value at (x_j, y_j) of node j and (x_m, y_m) of node m. This is a characteristic of an element shape function.

Alternatively, for the linear three node triangle, the element shape functions can be simply defined by area coordinates. Considering Figure 3.4, the coordinates L_1, L_2 and L_3 are defined as

$$L_1 = \frac{A_1}{A}$$

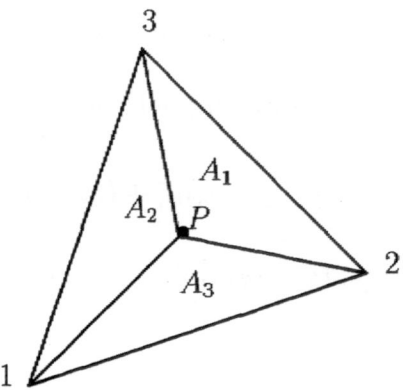

Figure 3.4: Area coordinate for triangular element

$$L_2 = \frac{A_2}{A}$$

$$L_3 = \frac{A_3}{A} \tag{3.74}$$

where A is the area $\triangle 123$, A_1 the area $\triangle P23$, A_2 the area $\triangle P31$ and A_3 the area $\triangle P12$. Since P has only two Cartesian coordinates (x, y), it must be the case that the three coordinates (L_i) cannot be independent of each other. It is clear that the sum of the three area coordinates equals one, that is

$$L_1 + L_2 + L_3 = 1 \tag{3.75}$$

It is seen that L_i has a unit value at node i and zero value at other nodes with a linear variation in between. It is easily verified that the area coordinates L_1, L_2 and L_3 are identical to the linear shape functions N_1^e, N_2^e and N_3^e. The Cartesian coordinates (x, y) may also be expressed as

$$\begin{bmatrix} 1 \\ x \\ y \end{bmatrix} = \begin{bmatrix} 1 & 1 & 1 \\ x_1 & x_2 & x_3 \\ y_1 & y_2 & y_3 \end{bmatrix} \begin{bmatrix} L_1 \\ L_2 \\ L_3 \end{bmatrix} \tag{3.76}$$

The following formula is useful in evaluating integrals involved in the finite element procedure where triangular elements have been used, that is

$$\int_{\Omega} L_1^a L_2^b L_3^c d\Omega = \frac{a!b!c!}{(a+b+c+2)!} 2A \qquad (3.77)$$

The **K** matrix for an individual element is given by substituting Equation (3.72) into Equation (3.61), that is

$$K_{ij}^e = \int_{\Omega_e} \left[\frac{\partial N_i^e}{\partial x} \frac{\partial N_j^e}{\partial x} + \frac{\partial N_i^e}{\partial y} \frac{\partial N_j^e}{\partial y} \right] dx dy$$

$$= \int_{\Omega_e} \frac{1}{4A^2} [c_i c_j + d_i d_j] d\Omega = \frac{1}{4A} [c_i c_j + d_i d_j] \qquad (i,j=1,2,3)$$
$$(3.78)$$

and the complete element matrix may be written as

$$\mathbf{K}^e = \frac{1}{4A} \begin{bmatrix} c_1 c_1 + d_1 d_1 & c_1 c_2 + d_1 d_2 & c_1 c_3 + d_1 d_3 \\ c_2 c_1 + d_2 d_1 & c_2 c_2 + d_2 d_2 & c_2 c_3 + d_2 d_3 \\ c_3 c_1 + d_3 d_1 & c_3 c_2 + d_3 d_2 & c_3 c_3 + d_3 d_3 \end{bmatrix} \qquad (3.79)$$

It can be seen $\mathbf{K}^e$ is symmetrical. The element load vector is given as follows

$$\mathbf{f}^e = \int_{\Omega_e} \begin{bmatrix} N_i^e \\ N_j^e \\ N_m^e \end{bmatrix} f(x,y)^e d\Omega \qquad (3.80)$$

and if $f(x,y)$ can be assumed as a constant in each element, that is $f(x,y)^e = F$ then

$$\mathbf{f}^e = \frac{FA}{3} \begin{bmatrix} 1 \\ 1 \\ 1 \end{bmatrix} \qquad (3.81)$$

Now, we undertake a two-dimensional finite element analysis to demonstrate the process of assembly for the element matrices. The problem is a two dimensional flow over a circular cylinder between parallel plates as shown in Figure 3.5. We try to obtain its stream function by solving the Laplace equation (2.30) in the section 2.7. As an example, the dimension in the flow direction is given arbitrarily and the inlet velocity is considered as constant shown in Figure 3.5. By taking advantage of the symmetry, the quadrant $a - b - c - d - e$ is idealised by 11 three-noded triangular elements with the nodal and element numbering shown in Figure 3.6. A

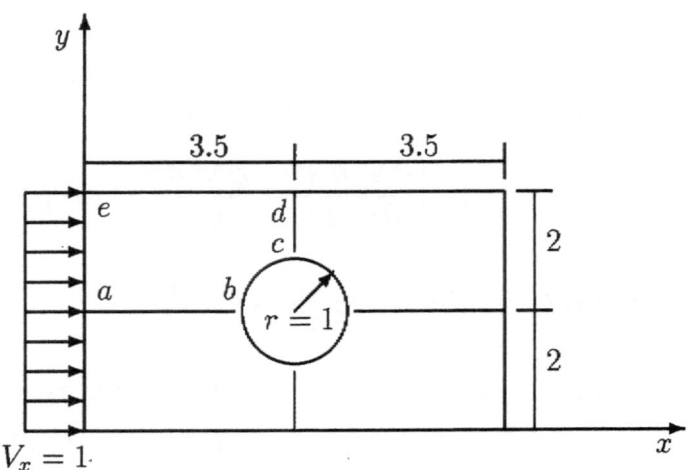

Figure 3.5: Flow around a cylinder.

coarse mesh is used for the demonstration purpose. The stream function at boundaries of $a-b$ and $b-c$ is set to be $\Psi = 0$ [3]. For the constant free stream velocity the Ψ varies linearly from 0 to 2 between $a-e$ and $\Psi = 2$ for boundary $d-e$.

The relationship between local and global numbering for each element is listed in Table 3.1 in an anticlockwise order.

From Equation (3.79), we obtain element stiffnesses

$$
\mathbf{K}^{(1)} = \mathbf{K}^{(2)} = \mathbf{K}^{(5)} = \mathbf{K}^{(6)} =
\begin{bmatrix}
2.6 & -2.5 & -0.1 \\
-2.5 & 2.5 & 0.0 \\
-0.1 & 0.0 & 0.1
\end{bmatrix}
\qquad (3.82)
$$

$$
\mathbf{K}^{(3)} = \mathbf{K}^{(4)} = \mathbf{K}^{(7)} = \mathbf{K}^{(8)} =
\begin{bmatrix}
1.25 & -1.0 & -0.25 \\
-1.0 & 1.0 & 0.0 \\
-0.25 & 0.0 & 0.25
\end{bmatrix}
\qquad (3.83)
$$

$$
\mathbf{K}^{(9)} = \mathbf{K}^{(10)} =
\begin{bmatrix}
1.45 & -1.25 & -0.2 \\
-1.25 & 1.25 & 0.0 \\
-0.2 & 0.0 & 0.2
\end{bmatrix}
\qquad (3.84)
$$

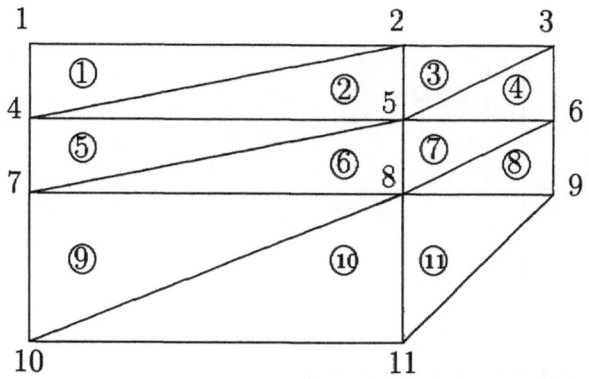

Figure 3.6: Finite element mesh.

Table 3.1: Element Connectivities

No. element	i	j	m	No. element	i	j	m
1	1	4	2	7	5	8	6
2	5	2	4	8	9	6	8
3	2	5	3	9	7	10	8
4	6	3	5	10	11	8	10
5	4	7	5	11	8	11	9
6	8	5	7				

and

$$\mathbf{K}^{(11)} = \begin{bmatrix} 1.0 & -0.5 & -0.5 \\ -0.5 & 0.5 & 0.0 \\ -0.5 & 0.0 & 0.5 \end{bmatrix} \tag{3.85}$$

Since, $\mathbf{f} = \mathbf{0}$ in Equations (3.80) and (3.81), we have

$$\mathbf{f} = \sum_{e=1}^{M} \mathbf{f}^e = 0 \tag{3.86}$$

According to equation (3.64), the global $\mathbf{K}$ matrix is assembled from the element matrices $\mathbf{K}^e$, thus

$$\mathbf{K} = \begin{bmatrix} K_{11} & K_{12} & & K_{14} & & & & & & & \\ & K_{22} & K_{23} & K_{24} & K_{25} & & & & & & \\ & & K_{33} & & K_{35} & K_{36} & & & & & \\ & & & K_{44} & K_{45} & & K_{47} & & & & \\ & & & & K_{55} & K_{56} & K_{57} & K_{58} & & & \\ & & & & & K_{66} & & K_{68} & K_{69} & & \\ & SYM. & & & & & K_{77} & K_{78} & & K_{710} & \\ & & & & & & & K_{88} & K_{89} & K_{810} & K_{811} \\ & & & & & & & & K_{99} & & K_{911} \\ & & & & & & & & & K_{1010} & K_{1011} \\ & & & & & & & & & & K_{1111} \end{bmatrix}$$

$$(3.87)$$

in which the non-zero terms are

$$K_{11} = K_{11}^1 \qquad K_{12} = K_{13}^1 \qquad K_{14} = K_{12}^1$$

$$K_{22} = K_{33}^1 + K_{22}^2 + K_{11}^3 \qquad K_{23} = K_{13}^3 \qquad K_{24} = K_{32}^1 + K_{23}^2$$
$$K_{25} = K_{21}^2 + K_{12}^3$$

$$K_{33} = K_{33}^3 + K_{22}^4 \qquad K_{35} = K_{32}^3 + K_{23}^4 \qquad K_{36} = K_{21}^4$$

$$K_{44} = K_{22}^1 + K_{33}^2 + K_{11}^5 \qquad K_{45} = K_{31}^2 + K_{13}^5 \qquad K_{47} = K_{12}^5$$

$$K_{55} = K_{11}^2 + K_{22}^3 + K_{33}^4 + K_{33}^5 + K_{22}^6 + K_{11}^7 \qquad K_{56} = K_{31}^4 + K_{13}^7$$
$$K_{57} = K_{32}^5 + K_{23}^6 \qquad K_{58} = K_{21}^6 + K_{12}^7$$

$$K_{66} = K_{11}^4 + K_{33}^7 + K_{22}^8 \qquad K_{68} = K_{32}^7 + K_{23}^8 \qquad K_{69} = K_{21}^8$$

$$K_{77} = K_{22}^5 + K_{33}^6 + K_{11}^9 \qquad K_{78} = K_{31}^6 + K_{13}^9 \qquad K_{710} = K_{12}^9$$

$$K_{88} = K_{11}^6 + K_{22}^7 + K_{33}^8 + K_{33}^9 + K_{22}^{10} + K_{11}^{11} \qquad K_{89} = K_{31}^8 + K_{13}^{11}$$
$$K_{810} = K_{32}^9 + K_{23}^{10} \qquad K_{811} = K_{21}^{10} + K_{12}^{11}$$

$$K_{99} = K_{11}^8 + K_{33}^{11} \qquad K_{911} = K_{32}^{11}$$

$$K_{1010} = K_{22}^9 + K_{33}^{10} \qquad K_{1011} = K_{31}^{10}$$

$$K_{1111} = K_{11}^{10} + K_{22}^{11}$$

It may be noted that the global $\mathbf{K}$ matrix is symmetrical about the leading diagonal and that it is also banded. The bandwidth of the matrix is affected by the method of numbering of the nodes. In general, the smaller the bandwidth, the greater the computational efficiency of the solution process. Now, the global $\mathbf{K}$ matrix may be written as

$$
\begin{bmatrix}
2.6 & -0.1 & 0 & -2.5 & 0 & 0 & 0 & 0 & 0 & 0 & 0 \\
-0.1 & 3.85 & -0.25 & 0 & -3.5 & 0 & 0 & 0 & 0 & 0 & 0 \\
0 & -0.25 & 1.25 & 0 & 0 & -1.0 & 0 & 0 & 0 & 0 & 0 \\
-2.5 & 0 & 0 & 5.2 & -0.2 & 0 & -2.5 & 0 & 0 & 0 & 0 \\
0 & -3.5 & 0 & -0.2 & 7.7 & -0.5 & 0 & -3.5 & 0 & 0 & 0 \\
0 & 0 & -1.0 & 0 & -0.5 & 2.5 & 0 & 0 & -1.0 & 0 & 0 \\
0 & 0 & 0 & -2.5 & 0 & 0 & 4.05 & -0.3 & 0 & -1.25 & 0 \\
0 & 0 & 0 & 0 & -3.5 & 0 & -0.3 & 6.3 & -0.75 & 0 & -1.75 \\
0 & 0 & 0 & 0 & 0 & -1.0 & 0 & -0.75 & 1.75 & 0 & 0 \\
0 & 0 & 0 & 0 & 0 & 0 & -1.25 & 0 & 0 & 1.45 & -0.2 \\
0 & 0 & 0 & 0 & 0 & 0 & 0 & -1.75 & 0 & -0.2 & 1.95
\end{bmatrix}
$$

From the boundary conditions we already know that

$$\Psi_9 = \Psi_{10} = \Psi_{11} = 0$$

and,

$$\Psi_1 = \Psi_2 = \Psi_3 = 2, \qquad \Psi_4 = 1.5, \qquad \Psi_7 = 1$$

Therefore, substituting the above known variables into the equation (3.5.3), the set of linear equations, in the matrix form, becomes

$$
\begin{bmatrix}
7.7 & -0.5 & -3.5 \\
-3.5 & 2.5 & 0 \\
-3.5 & 0 & 6.3
\end{bmatrix}
\begin{bmatrix}
\Psi_5 \\
\Psi_6 \\
\Psi_8
\end{bmatrix}
=
\begin{bmatrix}
7.3 \\
2 \\
0.3
\end{bmatrix}
$$

We can use the Gaussian elimination method to reduce a linear equation system to an upper triangular form by successive elimination. However, in present case, the solution can be easily worked out as

$$
\begin{aligned}
\Psi_5 &= 1.391 \\
\Psi_6 &= 1.078 \\
\Psi_8 &= 0.820
\end{aligned}
$$

For this coarse element mesh it is not worthwhile to compare the results with those obtained from analytical solution, however, the finite element results will approach the analytical solution when the mesh is further refined.

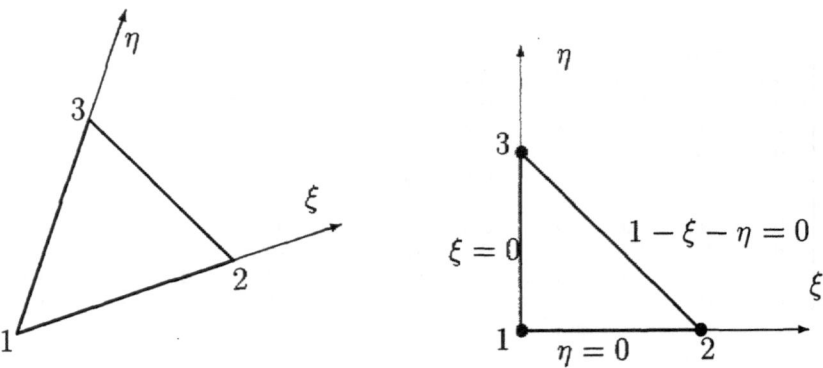

Figure 3.7: Mapping from general triangle to right-isosceles triangle

3.5.4   Natural coordinate system

There is another way to create shape functions for finite elements. As we know, a simple element, such as the square or the right-isosceles triangle can be mapped into a more complex shape in the global coordinate system.

Here, we consider the triangular mapping first. In a typical triangle (Figure 3.7), a skew coordinate system ξ, η is constructed by its side 1-2 and 1-3. The coordinates ξ, η at nodes 1, 2 and 3 are (0,0), (1,0) and (0,1) respectively. Therefore the actual element is transformed into a right-isosceles triangle in ξ, η which is called the master element. The functions defining three sides of the master element are given by $\eta = 0, 1 - \xi - \eta = 0$ and $\xi = 0$. Now, we can easily find out the shape functions as follows

$$N_1^e = 1 - \xi - \eta \qquad (3.88)$$

which has a unit value at node 1 and zero value at nodes 2, 3;

$$N_2^e = \xi \qquad (3.89)$$

which has a unit value at node 2 and zero value at nodes 3, 1;

$$N_3^e = \eta \qquad (3.90)$$

which has a unit value at node 3 and zero value at nodes 1, 2.

The global coordinate (x,y) can be expressed as

$$x = N_1^e x_1 + N_2^e x_2 + N_3^e x_3$$

$$y = N_1^e y_1 + N_2^e y_2 + N_3^e y_3 \qquad (3.91)$$

which is identical to Equation (3.76). In fact [4], here ξ is the same as L_2 and η is the same as L_3 and

$$N_1^e = 1 - \xi - \eta = L_1 \qquad (3.92)$$

In the derivation of the element **K** matrix (3.61), it has been necessary to establish the shape function derivatives with respect to the (x,y) coordinates. Since by chain rule, we have

$$\frac{\partial N_i}{\partial \xi} = \frac{\partial N_i}{\partial x}\frac{\partial x}{\partial \xi} + \frac{\partial N_i}{\partial y}\frac{\partial y}{\partial \xi}$$

$$\frac{\partial N_i}{\partial \eta} = \frac{\partial N_i}{\partial x}\frac{\partial x}{\partial \eta} + \frac{\partial N_i}{\partial y}\frac{\partial y}{\partial \eta} \qquad (3.93)$$

Therefore, the required derivatives $\frac{\partial N_i}{\partial x}$ and $\frac{\partial N_i}{\partial y}$ can be obtained by inversion as

$$\begin{bmatrix} \frac{\partial N_i}{\partial x} \\ \frac{\partial N_i}{\partial y} \end{bmatrix} = \mathbf{J}^{-1} \begin{bmatrix} \frac{\partial N_i}{\partial \xi} \\ \frac{\partial N_i}{\partial \eta} \end{bmatrix} \qquad (3.94)$$

where

$$\mathbf{J} = \begin{bmatrix} \frac{\partial x}{\partial \xi} & \frac{\partial y}{\partial \xi} \\ \frac{\partial x}{\partial \eta} & \frac{\partial y}{\partial \eta} \end{bmatrix} \qquad (3.95)$$

This mapping technique is specially useful for high order elements. For instance, the six node triangular element shown in Figure 3.8 has shape functions as

$$\begin{aligned} N_1^e &= \eta(2\eta - 1) \\ N_2^e &= 4\eta(1 - \xi - \eta) \\ N_3^e &= (1 - \xi - \eta)(1 - 2\xi - 2\eta) \\ N_4^e &= 4\xi(1 - \xi - \eta) \\ N_5^e &= \xi(2\xi - 1) \\ N_6^e &= 4\xi\eta \end{aligned} \qquad (3.96)$$

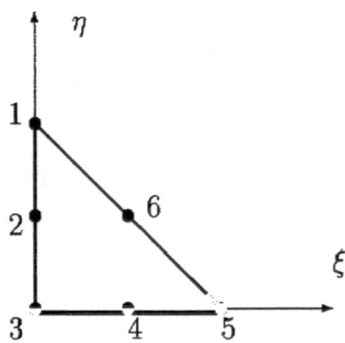

Figure 3.8: 6 node triangular element

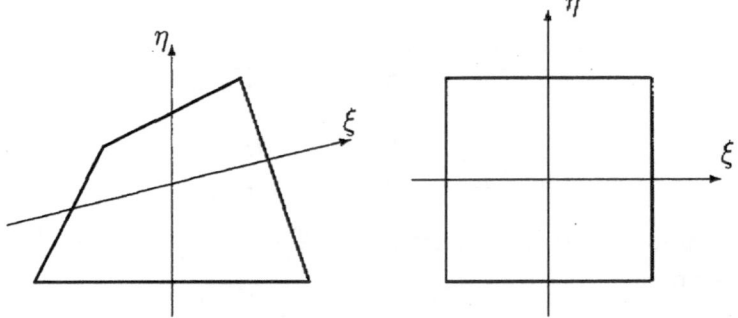

Figure 3.9: Mapping from quadrilateral to square

which are quadratic in (ξ, η).

In addition to triangles, quadrilateral elements are widely used. Figure 3.9 shows the mapping from a general quadrilateral to a square. The shape functions of some typical quadrilateral elements (Figure 3.10) are listed as follows.

Bilinear 4-node element

$$N_i^e = (1 + \xi\xi_i)(1 + \eta\eta_i) \tag{3.97}$$

Serendipity 8-node element

$$
\begin{aligned}
\text{Corner nodes: } N_i^e &= (1 + \xi\xi_i)(1 + \eta\eta_i)(\xi\xi_i + \eta\eta_i - 1)/4 \\
\text{Midside nodes: } N_i^e &= \xi_i^2(1 + \xi\xi_i)(1 - \eta^2)/2 \\
&\quad + \eta_i^2(1 + \eta\eta_i)(1 - \xi^2)/2
\end{aligned}
$$

Figure 3.10: 4,8 and 9 node elements

$$(3.98)$$

Lagrangian 9-node element

$$N_i^e = [\xi\xi_i(1+\xi\xi_i)/2+(1-\xi^2)(1-\xi_i^2)][\eta\eta_i(1+\eta\eta_i)/2+(1-\eta\eta^2)(1-\eta_i^2)]$$
$$(3.99)$$

where $\xi_i and \eta_i$ are the natural coordinates at the element node i.

3.5.5 Numerical integration

In order to evaluate the element $\mathbf{K}$ matrix, the following typical integral is involved (see 3.61).

$$I = \int_{\Omega_e} \left[\frac{\partial N_i^e}{\partial x} \frac{\partial N_j^e}{\partial x} \right] dxdy \qquad (3.100)$$

in which the element of area $dxdy$ has to be replaced by an equivalent in the ξ, η coordinates as

$$dxdy = det(\mathbf{J})d\xi d\eta \qquad (3.101)$$

Therefore Equation (3.100) can be rewritten in terms of integration over a square domain as

$$I = \int_{-1}^{1} \int_{-1}^{1} \left[\frac{\partial N_i^e}{\partial x} \frac{\partial N_j^e}{\partial x} \right] det(\mathbf{J})d\xi d\eta \qquad (3.102)$$

where $\frac{\partial N_i}{\partial x}^e, \frac{\partial N_j}{\partial x}^e$ are expressed in terms of ξ and η as given by Equation (3.94).

Other than for relatively simple cases (such as linear elements in triangles or bilinear elements in rectangles), the integrals involved in Equation (3.102) cannot be expressed in a simple analytic form. Therefore, numerical integration has to be used to evaluate such integrals. Here, Gauss-Legendre quadrature rules are adopted.

Numerical quadrature formulas in triangular elements have the form

$$I = \int_{\Omega} f(\xi, \eta) d\xi d\eta = \sum_{i=1}^{m} a_i f(\xi_i, \eta_i) \qquad (3.103)$$

where ξ_i and η_i are natural coordinates of the sampling points, a_i weighting factors and m the total number of integration points.

Numerical quadrature formulas in quadrilateral elements have the form

$$I = \int_{-1}^{1} \int_{-1}^{1} f(\xi, \eta) d\xi d\eta = \sum_{i=1}^{n} \sum_{j=1}^{n} a_i a_j f(\xi_i, \eta_j) \qquad (3.104)$$

where ξ_i, η_i are natural coordinates of the sampling points, a_i, a_j are weighting factors and n is number of integration points in one direction. The positions of the sampling points and the values of the weighting factors for both triangular and quadrilateral elements can be found in [1] and [5].

Bibliography

[1] O.C.Zienkiewicz and K.Morgan. *Finite Elements and Approximation*. John Wiley and Sons, New York, U.S.A., 1983.

[2] D.R.J.Owen and E.Hinton. *A Simple Guide to Finite Elements*. Pineridge Press, Swansea, U.K., 1980.

[3] T.J.Chung. *Finite Element Analysis in Fluid Dynamics*. McGRAW-HILL, New York, U.S.A., 1978.

[4] G.Carey and J.T.Oden. *Finite Elements, Vol II*. Pineridge Press, Englewood Cliffs, New Jersey 07632, 1983.

[5] H.C.Huang and A.S.Usmani. *The Finite Element Analysis for Heat Transfer*. Springer-Verlag, London, 1994.

Chapter 4

Steady Non-Newtonian Flow

4.1   Introduction

Steady non-Newtonian flow occurs in many situations. In the following context, a laminar behaviour has been discussed. It is known that in a laminar flow individual particles of fluid follow paths which do not cross those of neighbouring particles. This occurs when velocities are low enough for force due to viscosity to predominate over inertia forces. The relationship between inertial and viscous forces is given by the Reynolds number as follows

$$Re = \frac{\text{inertial forces}}{\text{viscous forces}} = \frac{\rho U L}{\mu} \qquad (4.1)$$

where U and L are suitable characteristic velocity and characteristic length. In most industrial practice, investigators are only interested in a fully developed flow regardless of the flow history. In that case, the velocities do not change with time, although the fluid at each position will possess different velocities.

Then the term

$$\frac{\partial}{\partial t} = 0 \qquad (4.2)$$

where the derivative operates on any parameters associated with flow. Therefore,

$$\frac{D}{Dt} = \frac{\partial}{\partial t} + \mathbf{v} \cdot \nabla = \mathbf{v} \cdot \nabla$$

which means that D/Dt is not equal to zero. $\mathbf{v} \cdot \nabla$ is the convection term which is due to the change of velocity in space. Steady flow can occur only if all the imposed conditions are constant in time. The momentum equation for steady flow becomes

$$\rho \mathbf{v} \cdot \nabla v = \rho \mathbf{g} + \nabla \cdot \sigma \qquad (4.3)$$

In the following lines we present a brief review of the main finite element techniques developed for steady incompressible flow problems [1].

The first finite element solutions of the incompressible Navier-Stokes equations, based on the *primitive variables* (velocity and pressure) as the nodal degree of freedoms, used an *integral* or *mixed* approach in which the nodal values were computed simultaneously in an iterative manner. This formulation resulted from recognizing that the incompressible flow equations are differential equations with applied Lagrangian constraints [2]. Such problems are characterised by two kinds of variables;

$$\begin{bmatrix} A & B \\ B^T & 0 \end{bmatrix} \begin{pmatrix} x \\ y \end{pmatrix} = \begin{pmatrix} f_1 \\ f_2 \end{pmatrix} \qquad (4.4)$$

where x are the primary variables and y are the constraint variables. Applied to the incompressible flow equations this results in a system of the form,

$$\begin{bmatrix} K & C \\ C^T & 0 \end{bmatrix} \begin{pmatrix} u \\ P \end{pmatrix} = \begin{pmatrix} f \\ 0 \end{pmatrix} \qquad (4.5)$$

where u and P are the velocities and pressures respectively, and P also corresponds to the Lagrange multipliers. When the standard Galerkin form of the finite element method (GFEM) is applied to such systems with an arbitrary combination of interpolation for pressures and velocities, the problem of mesh locking is encountered. It was discovered that a one order lower interpolation of pressure than that of the velocities was required to obtain viable solutions for GFEM formulations of incompressible flow. Two classes of elements were developed; Taylor-Hood elements (with continuous pressures) [3] and Crauzeix-Raviart elements (with discontinuous pressures) [4]. The elements that may be used for such systems are said to satisfy the *Babuska-Brezzi condition*. Further details on mixed formulations may be obtained from [5], and the mathematical theory can be found in reference [6]. The mixed method yields a system of equations which is not positive definite; special equation solvers are required as the diagonal terms corresponding to the continuity equation are zero. The computer storage and time requirements are generally large. In order to overcome some of these limitations, the penalty method was developed [7] which is an equivalent form of the mixed methods as shown by Malkus and Hughes [8]. The penalty method based formulation can be derived by expressing the Lagrange multipliers in terms of the primary variables;

$$P = \lambda f\left(u, v\right) \tag{4.6}$$

where λ is the penalty parameter. Malkus and Hughes [8] state that the element stiffness of the mixed method is identical to that of the penalty function method provided selective reduced integration has been used for the penalty terms. If selective reduced integration is not used then the problem of mesh locking is encountered. This is precisely what happens for the case of equal order of interpolation for pressures and velocities in the mixed formulation. At the reduced integration Gauss points, it is noted that [5, page 221] the pressure fields from the mixed method agree with the penalty term obtained using reduced integration, *i.e.*

$$P \approx -\lambda\left(\frac{\partial u}{\partial x} + \frac{\partial v}{\partial y}\right) \tag{4.7}$$

and this agreement improves as the value of λ increases. Therefore, λ is normally assigned as high a value as possible, which is limited by the floating point word length of the computer being used.

The application of the methods discussed above to the problem of steady non-Newtonian flow will be discussed in the following sections.

4.2   Creeping Motion

When Reynolds number is much smaller than unity the viscous forces dominate over the inertia forces. This special case is so-called *creeping motion* in which the convective term can either be neglected or considered as an additional part of body force [9] and we have

$$\nabla \cdot \sigma = \mathbf{b} \qquad (4.8)$$

where

$$\sigma_{ij} = -P\delta_{ij} + \tau_{ij}$$

and **b** are generalised body forces as follows

$$\mathbf{b} = \rho \mathbf{v}^* \cdot \nabla \mathbf{v}^* - \rho \mathbf{g} \qquad (4.9)$$

in which $\mathbf{v}^*$ are known velocities (for instance from a previous calculation).

The strain rate-velocity relations are

$$d_{ij} = \frac{1}{2}(u_{i,j} + u_{j,i}) \qquad (4.10)$$

with the continuity condition as follows

$$u_{i,i} = 0 \qquad (4.11)$$

There are two types of boundary conditions. The velocity boundary condition is as below

$$u_i = \bar{u}_i \qquad \text{on} \quad S_u \qquad (4.12)$$

in which $\bar{u}_i$ is the prescribed velocity. For the flow equations normal and tangential traction forces may be specified as boundary conditions, say, on S_f

$$\sigma_{ij} n_j = \bar{F}_{n_i} \qquad \text{on} \quad S_{fn} \qquad (4.13)$$

$$\sigma_{ij} t_j = \bar{F}_{t_i} \qquad \text{on} \quad S_{ft} \qquad (4.14)$$

where n and t are the unit normal and tangent vectors respectively to boundary S_f.

4.2.1 Extra stress and strain rate relations

The generalised constitutive law is applied here, that is

$$\boldsymbol{\tau} = 2\mu \mathbf{d} \tag{4.15}$$

where μ is considered to be dependent upon the second invariant of strain rate, such as (for a power-law fluid),

$$\mu(I_2) = \mu_0 I_2{}^{p-1} \tag{4.16}$$

where p is the power-law index, and μ_0 is the consistency factor.

As mentioned earlier (Section 2.3.3), there are two special cases of the equation (4.16). The Newtonian case is given by setting $p = 1$ (which leads to $\mu = \mu_0$). Setting $p = 0$ leads to,

$$\mu(I_2) = \frac{\mu_0}{I_2} \tag{4.17}$$

Identifying I_2 as the effective deviatoric strain-rate $\bar{\dot{\epsilon}}$ and redefining μ_0 as $\frac{1}{3}\bar{\sigma}$ ($\bar{\sigma}$ being the effective stress), we have,

$$\mu = \frac{1}{3} \frac{\bar{\sigma}}{\bar{\dot{\epsilon}}} \tag{4.18}$$

Which can be identified as the condition for perfect plastic flow obeying the Von-Mises yield criterion.

4.3 Finite Element Formulation for Creeping Flow

The virtual work principle (which is akin to the weighted residual method described in the previous chapter) will be used to achieve the spatial discretization of the equations governing creeping flow. The virtual work principle for steady state flow problems can be written as,

$$\int_\Omega [\delta \mathbf{d}]^T \, \sigma d\Omega + \int_\Omega [\delta \mathbf{v}]^T \, \mathbf{b} d\Omega = \int_{S_t} [\delta \mathbf{v}]^T \, \mathbf{t} dS_t \tag{4.19}$$

and the continuity part as,

$$\int_\Omega [\nabla \cdot \delta \mathbf{v}]^T (\lambda \nabla \cdot \mathbf{v} + \mathbf{P}) d\Omega = 0 \tag{4.20}$$

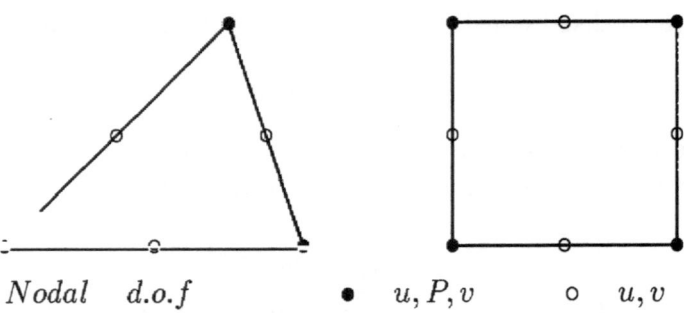

$$N odal \quad d.o.f \qquad \bullet \quad u, P, v \qquad \circ \quad u, v$$

Figure 4.1: Elements used for flow problems

where $\delta\mathbf{v}$ is the vector of virtual velocities, $\delta\mathbf{d}$ is the vector of associated virtual strain-rates, S_t is that part of the boundary on which boundary tractions are prescribed and $\mathbf{b}$ are the generalised body forces.

4.3.1   Finite element spatial discretisation

We now consider spatial discretisation in a two dimensional domain to obtain the finite element equations. For simplicity, we re-define stress and strain rate in the following vector forms.

$$\sigma = \left[\sigma_{11}, \sigma_{22}, \sigma_{33}, \sigma_{12}, \sigma_{23}, \sigma_{13}\right]^T$$

$$\boldsymbol{\tau} = \left[\boldsymbol{\tau}_{11}, \boldsymbol{\tau}_{22}, \boldsymbol{\tau}_{33}, \boldsymbol{\tau}_{12}, \boldsymbol{\tau}_{23}, \boldsymbol{\tau}_{13}\right]^T$$

$$\mathbf{d} = \left[d_{11}, d_{22}, d_{33}, d_{12}, d_{23}, d_{13}\right]^T \qquad (4.21)$$

where

$$\sigma = -\Theta^T P + \boldsymbol{\tau}$$

in which Θ is defined as

$$\Theta = [1, 1, 1, 0, 0, 0]$$

Elements with viable u and P interpolations are said to satisfy the Babuska-Brezzi [5, page 208] condition. Two commonly used elements that satisfy this condition are shown in Figure 4.1 (with *mixed* interpolation for velocity and pressure) and therefore may be used for spatial discretisation of the flow domain for our problem. These elements are not optimal in terms of the enforcement of the incompressibility condition, or, in other words, are relatively under-constrained, which leads to poor velocity solution on coarse grids and for difficult problems [10]. Special elements with better constraint properties are discussed in [10]. Here the elements of Figure 4.1 have been chosen for the reason of optimal convergence rate and a proven record of performance in problems of incompressible flow. The triangular element in the figure has been very strongly recommended by Zienkiewicz *et. al.* [11] for incompressible flow computations as well as non-Newtonian flow and elasto-plastic analysis.

The following shape functions will be used to approximate the field variables, in a typical element e

$$u_e = \sum_{i=1}^{n} N_i u_i$$

$$v_e = \sum_{i=1}^{n} N_i v_i \tag{4.22}$$

and as discussed earlier, a lower order interpolation for pressure,

$$P_e = \sum_{i=1}^{n'} M_i p_i \tag{4.23}$$

where n represents the number of velocity nodes in element e and n' the pressure nodes. And to reiterate, N represents quadratic shape functions for velocity interpolation and M represents linear shape functions for pressure interpolation.

Substitution of (4.22) and (4.23) into the virtual work equation (4.19) yields

$$\delta \mathbf{v}^T \left[\int_{\Omega} \mathbf{B}^T \sigma d\Omega - \int_{\Omega} \mathbf{N}^T \mathbf{b} d\Omega - \int_{S_t} \mathbf{N}^T \mathbf{t} dS_t \right] = 0 \tag{4.24}$$

where

$$\delta \mathbf{d} = \mathbf{B} \delta \mathbf{v}$$

and

$$B_{ij} = \frac{1}{2}(N_{i,j} + N_{j,i}) \tag{4.25}$$

Since $\delta\mathbf{v}$ is arbitrary, therefore we have

$$\int_\Omega \mathbf{B}^T \sigma d\Omega = \mathbf{f} \qquad (4.26)$$

where

$$\mathbf{f} = \int_\Omega \mathbf{N}^T \mathbf{b} + \int_{S_t} \mathbf{N}^T \mathbf{t} dS_t$$

from

$$\sigma = -\mathbf{\Theta}^T P + \boldsymbol{\tau}$$

we have

$$\int_\Omega \mathbf{B}^T \boldsymbol{\tau} d\Omega - \int_\Omega \mathbf{B}^T \mathbf{\Theta}^T P d\Omega = \mathbf{f} \qquad (4.27)$$

Using $\boldsymbol{\tau} = 2\mu\mathbf{d}$ and $P = \mathbf{M}p$, we obtain

$$\left(\int_\Omega \mathbf{B}^T 2\mu\mathbf{B} d\Omega\right)\mathbf{v} - \left(\int_\Omega \mathbf{B}^T \mathbf{\Theta}^\mathbf{T} \mathbf{M} d\Omega\right)\mathbf{p} = \mathbf{f} \qquad (4.28)$$

Substitution of (4.22) and (4.23) into the virtual work equation (4.20), using $\mathbf{N}\mathbf{v}$ for $\mathbf{v}$, yields

$$\delta\mathbf{v}^T \int_\Omega \left[\lambda \boldsymbol{\nabla} \cdot \mathbf{N}^\mathbf{T} \boldsymbol{\nabla} \cdot \mathbf{v} d\Omega + \boldsymbol{\nabla} \cdot \mathbf{N}^T P d\Omega\right] = 0 \qquad (4.29)$$

Since $\delta\mathbf{v}$ is arbitrary and using $\mathbf{M}\mathbf{p}$ for P, we have

$$\left(\int_\Omega \lambda \boldsymbol{\nabla} \cdot \mathbf{N}^\mathbf{T} \boldsymbol{\nabla} \cdot \mathbf{N} d\Omega\right)\mathbf{v} + \left(\int_\Omega \boldsymbol{\nabla} \cdot \mathbf{N}^\mathbf{T} \mathbf{M} d\Omega\right)\mathbf{p} = 0 \qquad (4.30)$$

It is noted that

$$\boldsymbol{\nabla} \cdot \mathbf{N} = \mathbf{\Theta}\mathbf{B} \qquad (4.31)$$

so

$$\left(\int_\Omega \lambda \mathbf{B}^\mathbf{T} \mathbf{\Theta}^\mathbf{T} \mathbf{\Theta} \mathbf{B} d\Omega\right)\mathbf{v} + \left(\int_\Omega \mathbf{B}^\mathbf{T} \mathbf{\Theta}^\mathbf{T} \mathbf{M} d\Omega\right)\mathbf{p} = 0 \qquad (4.32)$$

Combining both equations we have

$$\begin{bmatrix} \int_\Omega \mathbf{B}^T 2\mu\mathbf{B} d\Omega & \int_\Omega \mathbf{B}^T \mathbf{\Theta}^\mathbf{T} \mathbf{M} d\Omega \\ \int_\Omega \lambda \mathbf{B}^\mathbf{T} \mathbf{\Theta}^\mathbf{T} \mathbf{\Theta} \mathbf{B} d\Omega & \int_\Omega \mathbf{B}^\mathbf{T} \mathbf{\Theta}^\mathbf{T} \mathbf{M} d\Omega \end{bmatrix} \begin{pmatrix} \mathbf{v} \\ \mathbf{p} \end{pmatrix} = \begin{pmatrix} \mathbf{f} \\ 0 \end{pmatrix}$$

The above equations represent the mixed formulation of the Finite Element Method as defined in [2]. This form is widely used for the incompressible flow equations and for the steady state flow considered here, we have the following matrix form,

$$\begin{bmatrix} K_s & Q \\ Q^T & C \end{bmatrix} \begin{pmatrix} u \\ P \end{pmatrix} = \begin{pmatrix} f \\ 0 \end{pmatrix}$$

where,

$$\mathbf{K_s} = \int_\Omega \mathbf{B}^T 2\mu \mathbf{B} d\Omega)$$

$$\mathbf{Q} = \int_\Omega \mathbf{B}^T \mathbf{\Theta}^T \mathbf{M} d\Omega$$

$$\mathbf{C} = \int_\Omega \lambda \mathbf{B}^T \mathbf{\Theta}^T \mathbf{\Theta} \mathbf{B} d\Omega$$

As discussed earlier, the constraint variable p can be eliminated on element level using the equivalent *penalty formulation*, thus

$$\left[K_s + QC^{-1}Q^T \right]_e \bar{v}_e = f_e$$

$$\bar{p}_e = (C^{-1}Q^T)_e \bar{v}_e$$

After assembly, the global equation becomes

$$\mathbf{K}\bar{\mathbf{v}} = \mathbf{F} \tag{4.33}$$

where v are velocities F are total forces including generalised body forces. It may again be mentioned that all terms involving λ must be evaluated using reduced integration.

A finite element code, NSTEAD has been developed using the formulation described. The program can solve steady non-Newtonian problems with low Reynolds number such as polymer flow and metal forming. In some problems, slippage and friction conditions must also be considered. The following section presents a brief review of the various friction models used.

4.3.2 Slippage and friction

In non-Newtonian fluid flow, there might exist relative slippage between the boundary and the fluid, *e.g.* when squeezing tooth-paste in its tube. The relative movement produces friction on the flow boundary. Such friction may be calculated using the following equation according to the general friction law [12],

$$\mathbf{f}_F = -\alpha(\mathbf{v}_r, \sigma_n, \bar{\sigma}) \frac{\mathbf{v}_r}{|\mathbf{v}_r|} \tag{4.34}$$

where $\mathbf{f}_F$ is the friction force, $\mathbf{v}_r$ is the relative velocity along the boundary and α is the function of the relative velocity $\mathbf{v}_r$, the normal stress σ_n and the effective stress $\bar{\sigma}$. The friction force, $\mathbf{f}_F$, is always in the direction opposite the relative velocity and $\frac{\mathbf{v}_r}{|\mathbf{v}_r|}$ is the unit vector representing the direction of the relative velocity.

Tresca friction law:

For Tresca friction law we have

$$\alpha = \omega\tau \tag{4.35}$$

where ω is the friction factor and τ is the yield shear stress. Therefore Equation (4.34) becomes,

$$\mathbf{f}_F = -\omega\tau \frac{\mathbf{v}_r}{|\mathbf{v}_r|} \tag{4.36}$$

Coulomb friction law:

The Coulomb friction law is given by,

$$\alpha = \mu_F \sigma_\mathbf{n} \tag{4.37}$$

where μ_F is Coulomb friction coefficient, leading to,

$$\mathbf{f}_F = -\mu_F |\sigma_\mathbf{n}| \frac{\mathbf{v}_r}{|\mathbf{v}_r|} \tag{4.38}$$

Viscoplastic friction law:

The viscoplastic friction law is written as,

$$\alpha = \alpha_0 \mathbf{v}_r{}^q \tag{4.39}$$

where α_0 is a material constant and q is an experimentally determined parameter, usually taken from 0 to 1. Equation (4.34) for this law is,

$$\mathbf{f}_F = -\alpha_0 \frac{\mathbf{v}_r}{|\mathbf{v}_r|^{1-q}} \tag{4.40}$$

Only the viscoplastic friction law has been used in this text. The virtual work done by friction contributes to total virtual work. Therefore, Equation (4.33) is rewritten as,

$$\mathbf{K}\bar{\mathbf{v}} = \mathbf{F} + \int_S \mathbf{N^T f_F} dS \qquad (4.41)$$

Since $\mathbf{v}_r$ in most cases is the velocity at the boundaries, that is

$$\mathbf{v}_r = \mathbf{v} \qquad \text{on} \quad S_u \qquad (4.42)$$

we redefine

$$\mathbf{v}_r = \mathbf{v}_b \qquad (4.43)$$

and substituting (4.40) into (4.41) we have

$$\mathbf{K}\bar{\mathbf{v}} + \int_S \mathbf{N}^T \alpha \frac{\mathbf{v}}{|\mathbf{v}_b|} dS = \mathbf{F} \qquad (4.44)$$

or

$$(\mathbf{K} + \int_S \mathbf{N}^T \frac{\alpha}{|\mathbf{v}_b|} \mathbf{N} dS)\bar{\mathbf{v}} = \mathbf{F} \qquad (4.45)$$

where it is assumed that the friction term will disappear when $|\mathbf{v}_b|$ tends to zero.

4.3.3 Numerical examples

Two illustrative numerical examples are discussed in this section to demonstrate the application of the steady flow finite element program NSTEAD. These involve the simulation of flow over a step in a plane channel and a contraction tubular flow, which is an axisymmetric case. For these examples we use the six noded quadratic triangular elements with quadratic velocity components and linear pressures.

Example 1: Flow over a step.

Figure 4.2.a illustrates the geometry of the problem and the boundary conditions used in the finite element simulation. A unit uniform inlet flow is applied at left hand side as shown in the figure. The consistency factor in equation (4.16), μ_o=1.0, is used. The finite

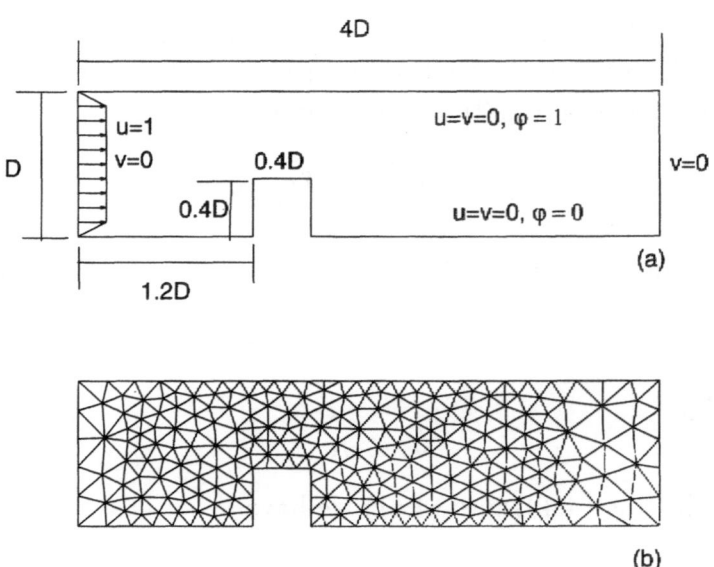

Figure 4.2: Flow over a step, showing, a) geometry and boundary conditions; b) finite element mesh

element mesh of six noded triangular elements is shown in Figure 4.2.b. The calculated results are plotted in Figure 4.3 for the case of Newtonian flow. The stream lines shown in Figure 4.3.a are drawn in 0.1 increment from zero at the lower boundary to unity at the upper boundary where values for the stream function are given. The velocity profiles of u in the x direction are plotted in part (b) of the figure which gives a maximum velocity of 2.3235 just above the step. It may be noted that the velocity profile is fully developed at the exit. Figure 4.3.c shows the flow direction, here the vectors are not proportional to the velocity magnitude but in a uniform size only to show the changes of the flow direction.

Further runs of this example were made for non-Newtonian flows with power index $p=0.5$ and 0.1. The results for these calculations are shown in Figures 4.4 and 4.5. It is clear that with decrease of power law index p the velocity profiles at the exit tend to be flattened, as we already observed in Figure 2.1 of chapter 2. The maximum velocity just above the step also reduced to 2.1281 and 1.9619, respectively.

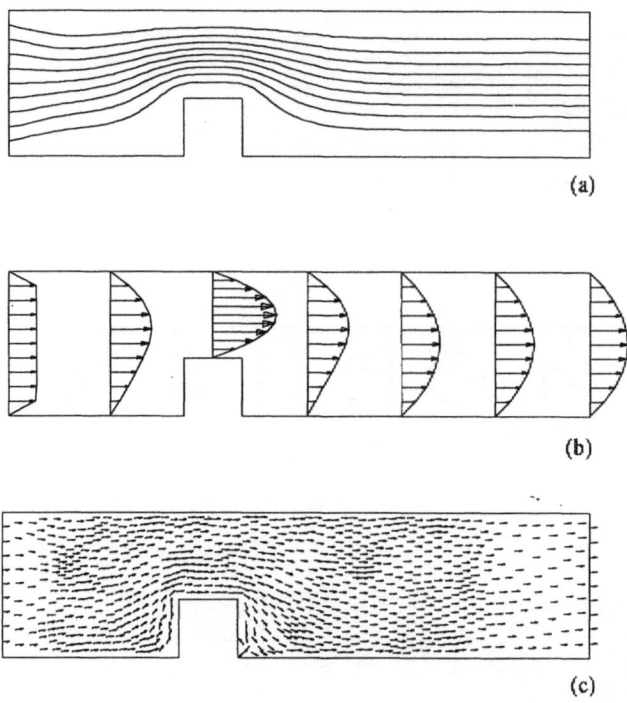

Figure 4.3: Flow over a step. Finite element results for Newtonian flow: a) stream lines; b) velocity profiles of u; c) fluid flow directions.

Example 2: Entry flow in a tubular contraction.

Figure 4.6 shows the geometry of the problem and the boundary conditions used, together with the finite element mesh. An axisymmetric 4/1 contraction is considered, where 4 is the ratio of the radius of the up stream tube to the one of the down stream tube. The upper stream has a length of 12 units and the down stream, 6 units. The boundary ABCD is a fixed wall on which the viscous fluid is assumed to stick, while EF is an axis of symmetry. A parabolic entry flow is imposed with a normalized maximum value of unity for $\bar{u}_z$ at the symmetry boundary. As it is expected that high velocities pressure gradients will occur in the neighbourhood of the contraction entrance, therefore the entrance and the contraction region are discretised using small elements.

We consider the flow of a power-law fluid through the 4/1 ax-

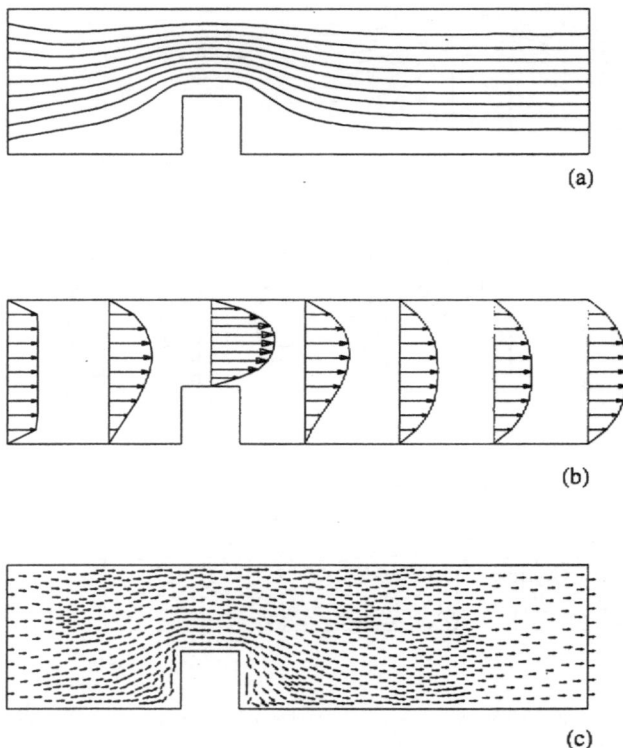

Figure 4.4: Flow over a step. Finite element results for p=0.5: a) stream lines; b) velocity profiles of u; c) fluid flow directions.

isymmetric contraction. The calculations were performed for power-law indices of 1.0, 0.75 and 0.0. Figure 4.7 shows the velocity profiles across the tube for all power index values considered. It may be seen from the figure that the final velocity is reached at the entrance to the small tube, and that the developed velocity profile is attained within a fraction of the down stream tube radius. The developed velocity profiles show very well the relation of the stream shape to the power index magnitude. Figure 4.8 shows the variation of the flow direction, and as before the vectors are not proportional to the velocity magnitude. It can also be seen that the corner vortex is only formed for the case of power index equals 1.0 (it is absent even for p=0.75). The region of still particles (where

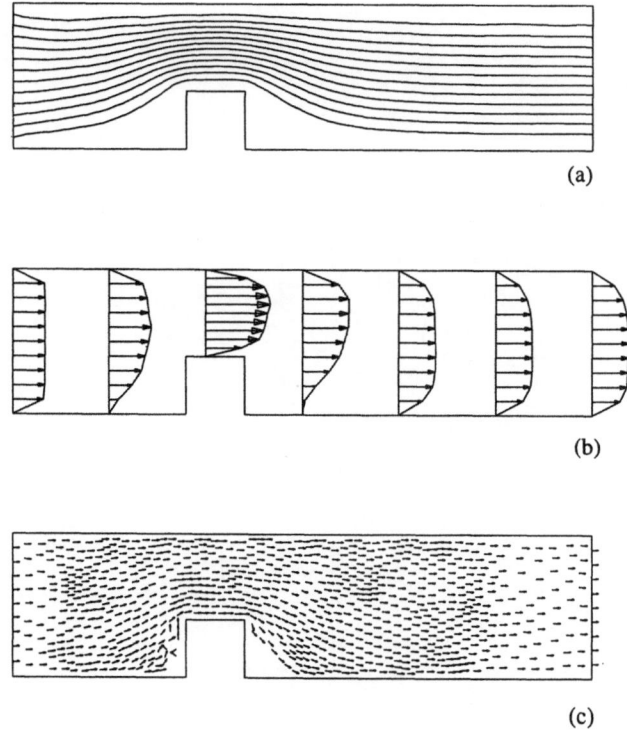

(a)

(b)

(c)

Figure 4.5: Flow over a step. Finite element results for p=0.1: a) stream lines; b) velocity profiles of u; c) fluid flow directions.

the flow vectors disappear), increases with decrease of the power index value. The axial velocity along the axis of symmetry is given in Figure 4.9, where r_o is the tube radius and $\bar{v}$ the maximum entrance flow at the axis of symmetry. Convergence is achieved for Newtonian flow within a few iterations, however, for smaller power indices many more iterations are needed. For the case of power index equals zero, it took about 50 iterations to reach the convergence tolerance required.

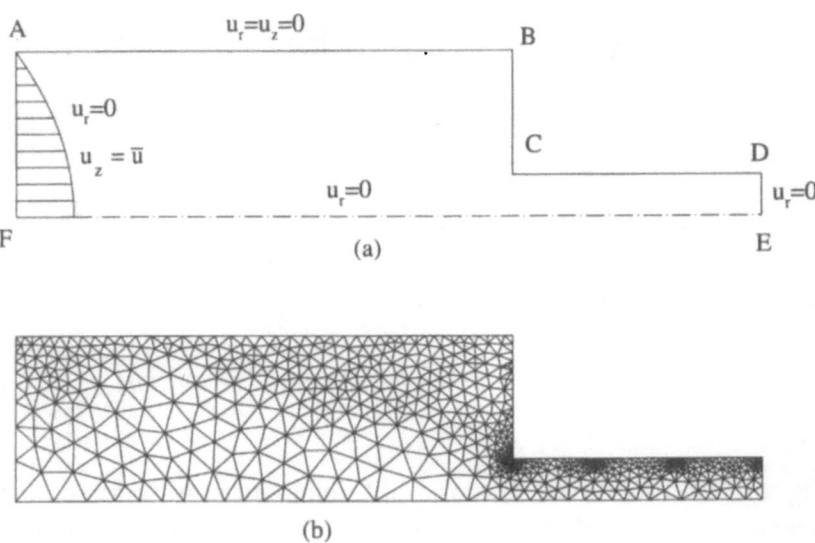

Figure 4.6: Tubular contraction flow. a) geometry and boundary conditions; b) finite element mesh for a 4/1 contraction flow.

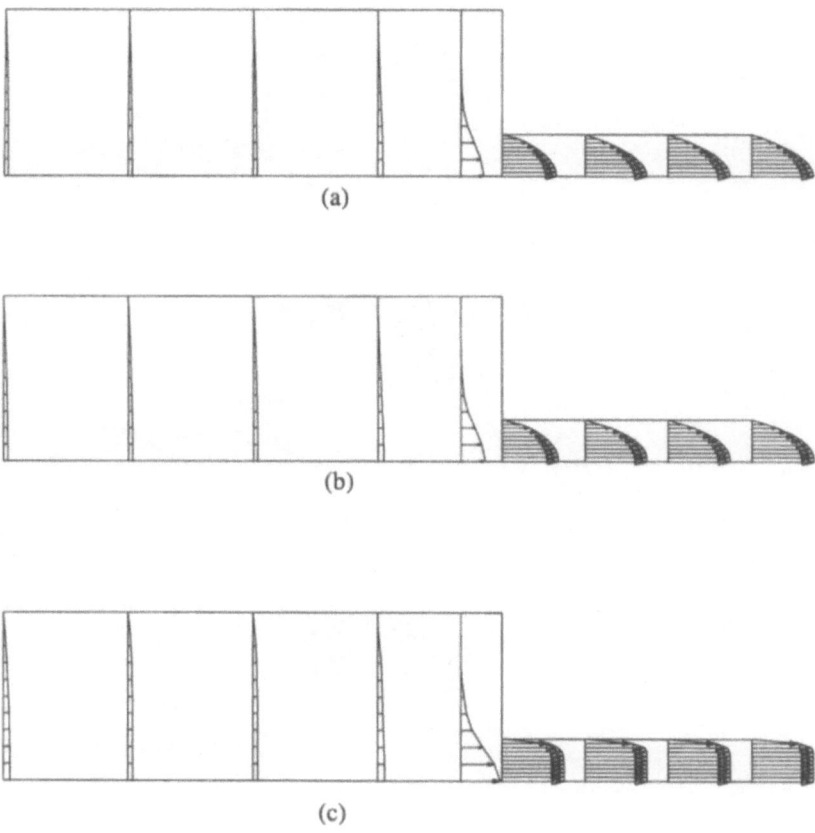

Figure 4.7: Development of axial velocity profiles across the tube:
a) p=1.0; b) p=0.75 c) p=0.0.

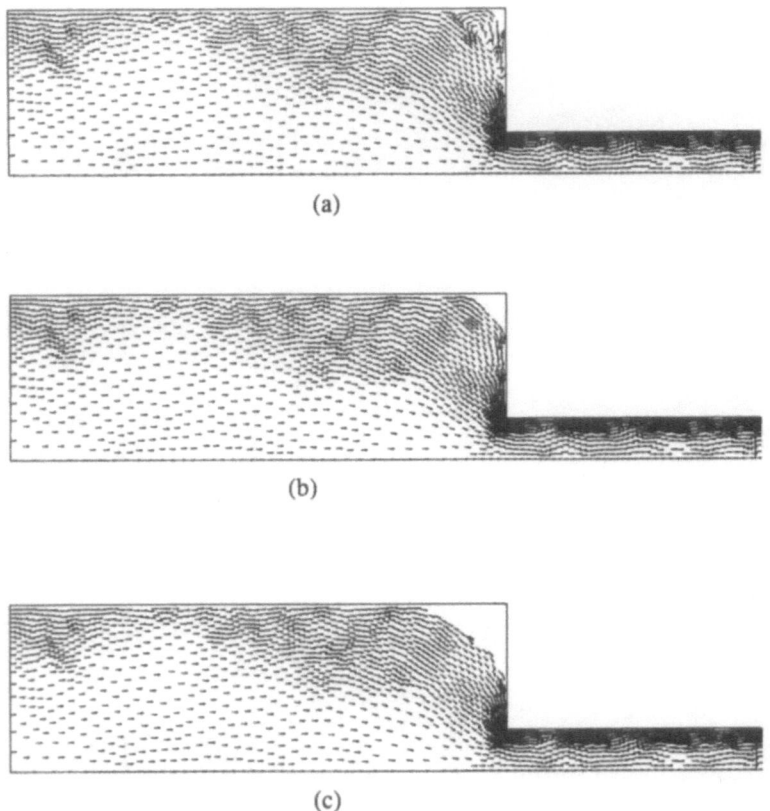

Figure 4.8: Variation of flow directions in the tube: a) p=1.0; b) p=0.75; c) p=0.0.

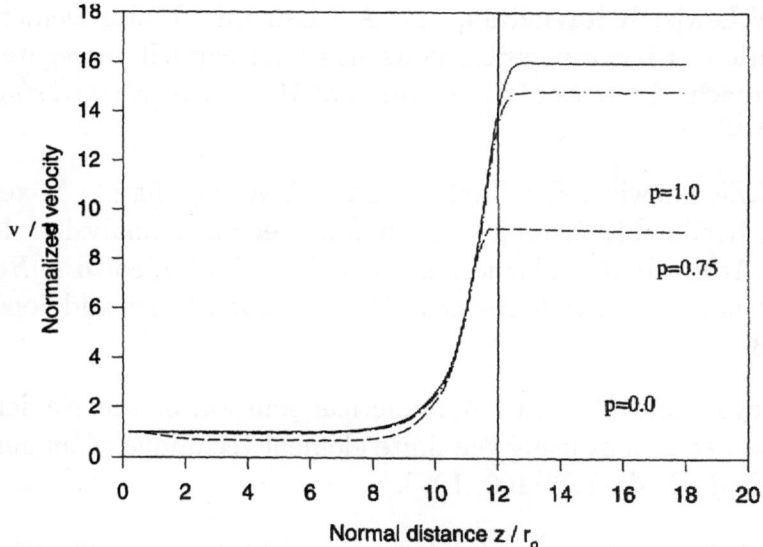

Figure 4.9: Comparison of the distribution of the axial velocity along the symmetry.

Bibliography

[1] R.W.Lewis, K.Ravindran, and A.S.Usmani. Finite element solution of incompressible flows using an explicit segregated approach. *Archives of Computational Methods in Engineering*, 2:69–93, 1995.

[2] O.C.Zienkiewicz, R.L.Taylor, and J.A.W.Baynham. Mixed and irreducible formulations in finite element analysis. In S.N.Atluri, R.H.Gallagher, and O.C.Zienkiewicz, editors, *Hybrid and mixed Finite Element Methods*, John Wiley and Sons, 1983.

[3] C.Taylor and P.Hood. A numerical solution of the Navier-Stokes equations using the finite element technique. *Computers and Fluids*, 1:73–100, 1973.

[4] M.Crouzeix and P.A.Raviart. *Conforming and nonconforming finite element methods for the stationary Stokes equation*. Technical Report R3, RAIRO, 1973.

[5] T.J.R.Hughes. *The Finite Element Method - Linear Static and Dynamic Finite Element Analysis*. Prentice-Hall International, Inc., Englewood Cliffs, New Jersey 07632, 1987.

[6] J.T.Oden and G.F.Carey. *Finite Elements: Mathematical Aspects, Vol IV*. Prentice-Hall, Englewood Cliffs, N.J., 1984.

[7] O.C.Zienkiewicz and P.N.Godbole. Viscous incompressible flow with special reference to non-newtonian (plastic) flow. In J.T.Oden, O.C.Zienkiewicz, R.H.Gallagher, and C.Taylor, editors, *Finite Element Methods in Fluids*, John Wiley and Sons, london, 1975.

[8] D.S.Malkus and T.J.R.Hughes. Mixed finite element methods–reduced and selective integration techniques: a unification of concepts. *Computer Methods in Applied Mechanics and Engineering*, 15:63–81, 1978.

[9] C.Taylor and T.G.Hughes. *Finite Element Programming of the Navier-Stokes Equations.* Pineridge Press, Swansea, U.K., 1981.

[10] P.M.Gresho, R.L.Lee, and R.L.Sani. On the time-dependent solution of the incompressible Navier-Stokes equations in two and three dimensions. In *Recent Advances in Numerical Methods in Fluids*, Pineridge Press Limited, Swansea, 1980.

[11] O.C.Zienkiewicz, Y.C.Liu, and G.C.Huang. Error estimates and convergence rates for various incompressible elements. *International Journal for Numerical Methods in Engineering*, 28:2191–2202, 1989.

[12] G.C.Huang. *Error Estimates and Adaptive Remeshing in Finite Element Analysis of Forming Processes - Ph.D. Thesis.* University of Wales, Swansea, 1989.

Chapter 5

Temporal Discretisation

5.1 Introduction

As opposed to *steady state* flow problems where velocities remains constant at a given point in the domain for all times, most fluid flow problems are time-dependent and referred to as *transient* flow problems.

The finite element discretisations discussed in the previous chapter were limited to the flow equations without the term containing the temporal derivative. Although, most real life flow problems are time-dependent, for many engineering problems it is sufficient to calculate a steady spatial velocity field. In fact, in many practical examples engineers are only interested in a steady state solutions. However, there are other problems where the transient effects cannot be ignored. For example, it may be required to calculate the history of the velocity and strain fields in metal forming processes. For such problems the complete flow equations including the temporal derivative term must be solved. Therefore, a temporal discretisation of the transient flow equations is required in addition to the spatial discretisation.

In the following sections we will present additional techniques required for the finite element analysis of transient problems. Apart from the basic temporal discretisation methods we will discuss other relevant topics such as automatic timestep selection.

5.1.1   Transient Newtonian flow equations in two dimensions

In order to demonstrate temporal discretisation, we use the Navier-Stokes equations for Newtonian flow in two dimensions as an example, *i.e.*:

The momentum equations,

$$\rho\left(\frac{\partial u}{\partial t} + u\frac{\partial u}{\partial x} + v\frac{\partial u}{\partial y}\right) = -\frac{\partial P}{\partial x} + \mu\left(2\frac{\partial^2 u}{\partial x^2} + \frac{\partial^2 u}{\partial y^2} + \frac{\partial^2 v}{\partial x \partial y}\right) \quad (5.1)$$

$$\rho\left(\frac{\partial v}{\partial t} + u\frac{\partial v}{\partial x} + v\frac{\partial v}{\partial y}\right) = -\frac{\partial P}{\partial y} + \mu\left(\frac{\partial^2 v}{\partial x^2} + 2\frac{\partial^2 v}{\partial y^2} + \frac{\partial^2 u}{\partial y \partial x}\right) \quad (5.2)$$

and the continuity equation,

$$\frac{\partial u}{\partial x} + \frac{\partial v}{\partial y} = 0 \quad (5.3)$$

5.1.2   Boundary conditions

The boundary conditions relevant to the class of problems considered are as follows:

(a) **Dirichlet or essential boundary conditions.**

These are applicable to the Navier Stokes equations as specified velocities at the boundaries. These may be constant or be allowed to vary with time, *i.e.*

$$\mathbf{v} = f(x, y, t) \qquad \text{on } S_{\mathbf{v}} \quad (5.4)$$

Pressure may not be specified at the boundaries as it is an implicit variable in an incompressible flow [1] which 'adjusts' itself to deliver a solenoidal (divergence free) velocity field. However, in the case of contained flow (prescribed velocities on all boundaries) the pressure becomes indeterminate and it must be specified at least at one point as a datum.

(b) **Neumann or natural boundary conditions.**

For the Navier Stokes equations normal and tangential traction forces may be specified as below, say, on S_f

$$f_n = -P + 2\mu \frac{\partial u_n}{\partial n} \tag{5.5}$$

$$f_\tau = \mu \left(\frac{\partial u_n}{\partial \tau} + \frac{\partial u_\tau}{\partial n} \right) \tag{5.6}$$

n and τ are the unit normal and tangent vectors respectively to boundary S_f.

S_v and S_f are portions of the boundary S of the computational domain Ω, in a way that the following relations hold,

$$S_v \cup S_f = S$$

$$S_v \cap S_f = \phi$$

where, ϕ is the null set.

5.2 Finite Element Formulation

The equations summarized in the previous section must first be discretised spatially to obtain the finite element equations. The conventional Galerkin weighted residual technique discussed in Chapter 3 (using the Poisson equation) is the most powerful and general method available to achieve finite element spatial discretisation for **any** set of differential equations. To demonstrate this let us consider the differential equation set represented by,

$$A(\phi) - f = 0 \quad \text{in} \quad \Omega \tag{5.7}$$

subject to certain boundary conditions on S. Where ϕ represents the unknown function and f is the known function of independent space coordinates. In the finite element method the continuous functions ϕ are approximated by piecewise shape or basis functions over small subdomains *i.e.*, elements, as,

$$\hat{\phi}_e = \sum_{i=1}^{n} N_i a_i \tag{5.8}$$

where $\hat{\phi}_e$ is the approximate value of ϕ over the element e, as interpolated from n discrete points on the element (nodes), according to the nodal parameters a_i and shape functions N_i. If we substitute the approximation $\hat{\phi}$ for the unknown function ϕ, we find that,

$$A(\hat{\phi}) - f = R \neq 0 \qquad (5.9)$$

where R is the residual. To reduce this residual to zero we use some trial or weighting function W *i.e.*,

$$\int_{\Omega} W[A(\hat{\phi}) - f]d\Omega = 0 \qquad (5.10)$$

If the approximating shape functions N_i, satisfy the boundary conditions of (5.7) and are of sufficient order to enable the integration of (5.7) (which is generally one less than that of the operator A), then absolutely any function can be chosen for W. In normal finite element practice W is chosen to be the same as the approximating functions N. This is known as the Galerkin or more precisely Bubnov-Galerkin form of the weighted residual method. The choices of $W \neq N$ are collectively termed as Petrov-Galerkin.

Sometimes the order of operator A may require higher order shape functions for interpolating the field variable within the element. This may be reduced by one by applying Greens theorem to equation (5.10), which produces,

$$\int_{\Omega} W A(\hat{\phi})d\Omega = \int_{\Omega} C(W)B(\hat{\phi})d\Omega + \int_{S} E(W)D(\hat{\phi})dS \qquad (5.11)$$

The boundary integral containing E and D corresponds to the natural boundary conditions for the problem. As discussed earlier (Chapter 3), Equation (5.11) is called the 'weak' form of the original problem, where B,C,D and E are all one order lower than A. We note here that the weighting functions W in the weak form are acted upon by the operator C which imposes on them continuity restrictions of one order higher than before, which, however, does not create any difficulty, since we use the same interpolation functions for W as for ϕ according to the Galerkin procedure.

Applying now the Galerkin weighted residual procedure to our differential equations with a mixed formulation using following shape functions (as in Chapter 4),

$$\text{Velocities} \qquad u_e = \sum_{i=1}^{n} N_i u_i$$

$$v_e = \sum_{i=1}^{n} N_i v_i$$

$$\text{Pressure} \qquad P_e = \sum_{i=1}^{n'} N_i' P_i \qquad (5.12)$$

we obtain the equations very popular and successful Galerkin finite element method (GFEM) equations discretised in space [1]. This is illustrated in the following lines.

If we approximate (5.1), (5.2) and (5.3) and the boundary conditions (5.5) and (5.6) using the shape functions in (5.12) then the equations are not exactly satisfied and a residual is obtained. This residual is then minimized using weighting functions W. The following equations are obtained.

Continuity

$$\int_{\Omega} W_i \left(\frac{\partial N_j}{\partial x} u_j + \frac{\partial N_j}{\partial y} v_j \right) d\Omega = 0 \qquad (5.13)$$

x-Momentum

$$\int_{\Omega} W_i \rho \left(N_j \dot{u}_j + N_k u_k \frac{\partial N_j}{\partial x} u_j + N_k v_k \frac{\partial N_j}{\partial y} u_j \right) d\Omega$$

$$+ \int_{\Omega} W_i \left(\frac{\partial N_j'}{\partial x} P_j \right) d\Omega$$

$$- \int_{\Omega} W_i \mu \left(2 \frac{\partial^2 N_j}{\partial x^2} u_j + \frac{\partial^2 N_j}{\partial y^2} u_j + \frac{\partial^2 N_j}{\partial x \partial y} v_j \right) d\Omega$$

$$= 0 \qquad (5.14)$$

y-Momentum

$$\int_\Omega W_i \rho \left(N_j \dot{v}_j + N_k u_k \frac{\partial N_j}{\partial x} v_j + N_k v_k \frac{\partial N_j}{\partial y} v_j \right) d\Omega$$

$$+ \int_\Omega W_i \left(\frac{\partial N_j'}{\partial y} P_j \right) d\Omega$$

$$- \int_\Omega W_i \mu \left(\frac{\partial^2 N_j}{\partial x^2} v_j + 2 \frac{\partial^2 N_j}{\partial y^2} v_j + \frac{\partial^2 N_j}{\partial x \partial y} u_j \right) d\Omega$$

$$= 0 \tag{5.15}$$

Before considering the boundary conditions, Green's theorem is applied to the equations of Momentum and Energy to get rid of the second derivatives. We will also apply Green's theorem to the pressure term for reasons which will be apparent later. For illustration, the procedure is demonstrated only for the x-momentum equations. When Green's theorem is applied to the x-momentum equation we obtain,

$$\int_\Omega W_i \rho \left(N_j \dot{u}_j + N_k u_k \frac{\partial N_j}{\partial x} u_j + N_k v_k \frac{\partial N_j}{\partial y} u_j \right) d\Omega$$

$$- \int_\Omega \frac{\partial W_i}{\partial x} \left(N_j' P_j \right) d\Omega + \int_S W_i \left(N_j' P_j \right) l_x dS$$

$$+ \int_\Omega \mu \left(2 \frac{\partial W_i}{\partial x} \frac{\partial N_j}{\partial x} u_j + \frac{\partial W_i}{\partial y} \frac{\partial N_j}{\partial y} u_j + \frac{\partial W_i}{\partial y} \frac{\partial N_j}{\partial x} v_j \right) d\Omega$$

$$- \int_S W_i \mu \left[\left(2 \frac{\partial N_j}{\partial x} u_j \right) l_x + \left(\frac{\partial N_j}{\partial y} u_j + \frac{\partial N_j}{\partial x} v_j \right) l_y \right] dS$$

$$= 0 \tag{5.16}$$

Now consider the boundary conditions of normal and tangential tractions as given by (5.5) and (5.6). Writing the boundary conditions for x-direction and using the weighting functions, we obtain, for boundary S_f (where the tractions are specified),

$$- \int_{S_f} W_i \ f_x + \int_{S_f} W_i \left(-N_j' P_j + 2\mu \frac{\partial N_j}{\partial x} u_j \right) l_x dS$$

$$+ \int_{S_f} W_i \ \mu \left(\frac{\partial N_j}{\partial y} u_j + \frac{\partial N_j}{\partial x} v_j \right) l_y dS = 0 \tag{5.17}$$

Comparing (5.16) and (5.17) we find that the expression obtained for the natural boundary conditions in (5.17) is contained in (5.16) for the whole boundary S. Adding (5.17) to (5.16) we find that the traction boundary conditions are automatically satisfied upon S_f as the expressions cancel, which is why they are referred to as natural boundary conditions. For the remaining boundary *i.e.*, $S - S_f$, if we specify the variables u and v, *i.e.* apply forced boundary conditions, we can choose the weighting functions W to be zero on $S - S_f$ which results in the final form of the x-momentum equation. Using the Galerkin form, ie, $N_j = W_j$, we have,

$$\int_\Omega \rho \left(N_i N_j \dot{u}_j + N_i N_k u_k \frac{\partial N_j}{\partial x} u_j + N_i N_k v_k \frac{\partial N_j}{\partial y} u_j \right) d\Omega$$

$$- \int_\Omega \frac{\partial N_i}{\partial x} N_j' P_j d\Omega$$

$$+ \int_\Omega \mu \left(2 \frac{\partial N_i}{\partial x} \frac{\partial N_j}{\partial x} u_j + \frac{\partial N_i}{\partial y} \frac{\partial N_j}{\partial y} u_j + \frac{\partial N_i}{\partial y} \frac{\partial N_j}{\partial x} v_j \right) d\Omega$$

$$= \int_{S_f} N_i f_x \tag{5.18}$$

Using the same process, we get the final forms of the y-momentum equation and the energy equation. In the continuity equation, (5.13), we must use the pressure shape functions N_i' for the weighting functions as the continuity equation will only be enforced at the pressure modes.

We may write the final set of spatially discretised equations in a fully coupled form as follows,

$$\mathbf{M}\dot{\boldsymbol{\theta}} + \mathbf{K}\boldsymbol{\theta} = \mathbf{F} \tag{5.19}$$

Equation (5.19) may be written in an expanded matrix form as

$$\begin{bmatrix} \rho\mathbf{M}_u & 0 & 0 \\ 0 & 0 & 0 \\ 0 & 0 & \rho\mathbf{M}_v \end{bmatrix} \begin{pmatrix} \dot{\mathbf{u}} \\ \dot{\mathbf{P}} \\ \dot{\mathbf{v}} \end{pmatrix} + \begin{bmatrix} \mathbf{K}_{uu} & \mathbf{C}_u & \mathbf{K}_{uv} \\ \mathbf{C}_u^T & 0 & \mathbf{C}_v^T \\ \mathbf{K}_{vu} & \mathbf{C}_v & \mathbf{K}_{vv} \end{bmatrix} \begin{pmatrix} \mathbf{u} \\ \mathbf{P} \\ \mathbf{v} \end{pmatrix}$$

$$= \begin{pmatrix} \mathbf{F}_u \\ 0 \\ \mathbf{F}_v \end{pmatrix}$$

Where the first to the fourth rows represent the energy, x-momentum, continuity and y-momentum equation respectively. The matrix components are as follows,

$$\mathbf{M}_u = \mathbf{M}_v = \int_\Omega N_i N_j$$

All these matrices are $n \times n$, with n being the number of velocity interpolation nodes.

$$\mathbf{K}_{uu} = \int_\Omega \rho \left(N_i N_k u_k \frac{\partial N_j}{\partial x} + N_i N_k v_k \frac{\partial N_j}{\partial y} \right)$$
$$+ \mu \left(2 \frac{\partial N_i}{\partial x} \frac{\partial N_j}{\partial x} + \frac{\partial N_i}{\partial y} \frac{\partial N_j}{\partial y} \right)$$

$$\mathbf{K}_{uv} = \int_\Omega \mu \left(\frac{\partial N_i}{\partial y} \frac{\partial N_j}{\partial x} \right)$$

$$\mathbf{K}_{vv} = \int_\Omega \rho \left(N_i N_k u_k \frac{\partial N_j}{\partial x} + N_i N_k v_k \frac{\partial N_j}{\partial y} \right)$$
$$+ \mu \left(\frac{\partial N_i}{\partial x} \frac{\partial N_j}{\partial x} + 2 \frac{\partial N_i}{\partial y} \frac{\partial N_j}{\partial y} \right)$$

$$\mathbf{K}_{vu} = \int_\Omega \mu \left(\frac{\partial N_i}{\partial x} \frac{\partial N_j}{\partial y} \right)$$

All the above are $n \times n$ matrices.

$$\mathbf{C}_u = -\int_\Omega \frac{\partial N_i}{\partial x} N_j'$$
$$\mathbf{C}_v = -\int_\Omega \frac{\partial N_i}{\partial y} N_j'$$

These are $n \times n'$ matrices.

$$\mathbf{C}_u^T = \int_\Omega N_i' \frac{\partial N_j}{\partial x}$$
$$\mathbf{C}_v^T = \int_\Omega N_i' \frac{\partial N_j}{\partial y}$$

These are $n' \times n$ matrices.

Finally, the force vectors, which are all n vectors.

$$\mathbf{F}_u = \int_S N_i f_x$$
$$\mathbf{F}_v = \int_S N_i f_y$$

5.3 Temporal Discretisation

The spatial discretisation discussed in the previous section leaves us with the first order ordinary differential equations with respect to time. Again, there are numerous ways of accomplishing the discretisation of the time domain. One can write finite element shape functions to include the time variable and thus incorporate it into the general finite element method procedure [2]. However, due to the conceptual simplicity of the time dimension simpler finite difference approximations are generally favoured, and most schemes currently used are constructed in this way.

5.3.1 Generalized trapezoidal and mid-point family of methods

The first order system of equations represented by (5.19), needs to be discretised in time. The most commonly used method for such a system is the "Generalized mid-point or trapezoidal family of methods", see [3, Chapter 8] and [4, Chapter 2] for details. The trapezoidal method when applied to (5.19), can be written as follows,

$$\mathbf{M}(\boldsymbol{\theta}_{n+1}, t_{n+1})\,\dot{\boldsymbol{\theta}}_{n+1} + \mathbf{K}(\boldsymbol{\theta}_{n+1}, t_{n+1})\,\boldsymbol{\theta}_{n+1} = \mathbf{F}_{n+1} \qquad (5.20)$$

and

$$\frac{\dot{\boldsymbol{\theta}}_{n+1} + \dot{\boldsymbol{\theta}}_n}{2} = \frac{\boldsymbol{\theta}_{n+1} - \boldsymbol{\theta}n}{\Delta t_n} \qquad (5.21)$$

subscripts n represent the nth time step. If not indicated otherwise Δt will mean Δt_n. Substituting (5.21) in (5.20), we obtain,

$$\left[\frac{2\mathbf{M}_{n+1}}{\Delta t} + \mathbf{K}_{n+1}\right](\boldsymbol{\theta}_{n+1}) = [\mathbf{M}_{n+1}]\left(\frac{2}{\Delta t}\boldsymbol{\theta}_n + \dot{\boldsymbol{\theta}}_n\right) + (\mathbf{F}_{n+1})$$

$$(5.22)$$

This method involves the calculation of the derivatives on the right hand side. Here, $\mathbf{M}_{n+1}$ etc. mean $\mathbf{M}(\boldsymbol{\theta}_{n+1}, t_{n+1})$.

The generalized mid-point family of methods [4, page 145] is written as,

$$\mathbf{M}(\boldsymbol{\theta}_{n+\alpha}, t_{n+\alpha})\dot{\boldsymbol{\theta}}_{n+\alpha} + \mathbf{K}(\boldsymbol{\theta}_{n+\alpha}, t_{n+\alpha})\boldsymbol{\theta}_{n+\alpha} = \mathbf{F}(\boldsymbol{\theta}_{n+\alpha}, t_{n+\alpha}) \quad (5.23)$$

where,

$$\begin{aligned}
\boldsymbol{\theta}_{n+\alpha} &= (1-\alpha)\boldsymbol{\theta}_n + \alpha\boldsymbol{\theta}_{n+1} \\
\dot{\boldsymbol{\theta}}_{n+\alpha} &= \frac{\boldsymbol{\theta}_{n+1} - \boldsymbol{\theta}_n}{\Delta t} \\
t_{n+\alpha} &= t_n + \alpha\Delta t
\end{aligned} \qquad (5.24)$$

Substituting (5.24) into (5.23) we obtain,

$$\left[\frac{\mathbf{M}_{n+\alpha}}{\Delta t} + \alpha\mathbf{K}_{n+\alpha}\right](\boldsymbol{\theta}_{n+1}) = \left[\frac{\mathbf{M}_{n+\alpha}}{\Delta t} - (1-\alpha)\mathbf{K}_{n+\alpha}\right](\boldsymbol{\theta}_n) + (\mathbf{F}_{n+\alpha})$$

$$(5.25)$$

No calculation of derivatives is necessary for this method. By changing the values of α from 0 to 1, different members of this family of methods are identified, *i.e.*,

$\alpha = 0$ -Forward Difference or Forward Euler.

$\alpha = \frac{1}{2}$ -Midpoint rule or Crank Nicolson.

$\alpha = \frac{2}{3}$ -Galerkin.

$\alpha = 1$ -Backward Difference or Backward Euler.

All, except the first (forward Euler), of the above schemes are implicit, i.e., they require matrix inversion for solution.

The stability and convergence of non-linear convection diffusion problems, which involve non-symmetrical and non-positive definite matrices have been treated with relatively less rigour in the relevant literature such as [4]. Some results obtained from an energy method analysis [4, page 150] suggest unconditioned stability for

$\alpha \geq \frac{1}{2}$ for the generalized mid-point family, which may be extended to the trapezoidal rule as well, using the general equivalence of the trapezoidal and the mid-point family [5], as suggested by Hughes [3, page 150].

As far as accuracy is concerned the mid-point rule is to be preferred [6]. Also, Cliffe [7] has shown that the generalized mid-point rule conserves linear and quadratic quantities, while the trapezoidal rule conserves only the linear ones.

5.3.2   High order scheme - Taylor series in time

Generalized mid-point and trapezoidal family of methods are first order time discretisations which have lower accuracy for integration in time. To improve upon this, high order expansion of Taylor series in time may be used. To illustrate a high order scheme we consider the one-dimensional scalar conservation law represented by the following equations:

$$\frac{\partial u}{\partial t} + \frac{\partial f(u)}{\partial x} = 0$$

and

$$a(x) = \frac{\partial f}{\partial u} \tag{5.26}$$

where u is a function of t and x. Expanding the solution $u(x,t)$ in a Taylor series in the time-step Δt, correct to second order in time gives

$$u^{n+1} = u^n + \Delta t \left[\frac{\partial u}{\partial t}\right]^n + \frac{1}{2}\Delta t^2 \left[\frac{\partial^2 u}{\partial t^2}\right]^n \tag{5.27}$$

where n denotes the time level such that $t^n = n\Delta t$. Considering

$$\frac{\partial u}{\partial t} = -\frac{\partial f(u)}{\partial x}$$

thus

$$\begin{aligned} \frac{\partial^2 u}{\partial t^2} &= -\frac{\partial}{\partial t}\left(\frac{\partial f}{\partial x}\right) \\ &= -\frac{\partial}{\partial u}\left(\frac{\partial f}{\partial x}\right)\frac{\partial u}{\partial t} \end{aligned}$$

$$= \frac{\partial}{\partial x}\left(\frac{\partial f}{\partial u}\right)\frac{\partial f}{\partial x} \tag{5.28}$$

Therefore, we have

$$u^{n+1} = u^n + \Delta t\left[\frac{-\partial f}{\partial x}\right]^n + \frac{1}{2}\Delta t^2\left[\frac{\partial}{\partial x}\left(\frac{\partial f}{\partial u}\right)\frac{\partial f}{\partial x}\right]^n \tag{5.29}$$

Therefore it appears that in order to evaluate the u^{n+1} from u^n one has to calculate terms such as $\frac{\partial f}{\partial u}$. To avoid the explicit evaluation of $\frac{\partial f}{\partial u}$, let us reconsider Equation (5.29). From equation (5.27) we can rewrite u^{n+1} as

$$
\begin{aligned}
u^{n+1} &= u^n + \Delta t\left[-\frac{\partial f}{\partial x}\right]^n + \frac{1}{2}\Delta t^2 \cdot \left[-\frac{\partial}{\partial t}\left(\frac{\partial f}{\partial x}\right)\right]^n \\
&= u^n + \Delta t\left(\left[-\frac{\partial f}{\partial x}\right]^n + \frac{1}{2}\Delta t\left[-\frac{\partial}{\partial t}\left(\frac{\partial f}{\partial x}\right)\right]^n\right) \\
&= u^n + \Delta t\left[-\frac{\partial f}{\partial x}\right]^{n+\frac{1}{2}} \tag{5.30}
\end{aligned}
$$

where,

$$\left[-\frac{\partial f}{\partial x}\right]^{n+\frac{1}{2}} = \left[-\frac{\partial f(u^{n+\frac{1}{2}})}{\partial x}\right] \tag{5.31}$$

and we can easily calculate $u^{n+\frac{1}{2}}$ from the following equation.

$$
\begin{aligned}
u^{n+\frac{1}{2}} &= u^n + \frac{1}{2}\left[\frac{\partial u}{\partial t}\right]^{n+\frac{1}{2}} \\
&= u^n + \frac{1}{2}\Delta t\left(-\frac{\partial f(u^{n+\frac{1}{2}})}{\partial x}\right) \tag{5.32}
\end{aligned}
$$

5.3.3 Predictor-corrector scheme

Using Equations (5.30) and (5.32) with the one-dimensional scalar conservation equation one can obtain higher accuracy in time integration without extra computational efforts. This is achieved by employing a two-step approach as follows.

Step1 :

$$u^{n+\frac{1}{2}} = u^n + \frac{1}{2}\Delta t \left(-\frac{\partial f(u^n)}{\partial x}\right) \tag{5.33}$$

Step2 :

$$u^{n+1} = u^n + \Delta t \left(-\frac{\partial f(u^{n+\frac{1}{2}})}{\partial x}\right) \tag{5.34}$$

The equation in step 1 is the predictor and the equation in step 2 is the corrector.

We consider now the ordinary differential equations resulting from the GFEM spatial discretisation. The approximate solution $U(x,t)$ is introduced over a finite dimensional space of functions such that

$$U(x,t) = N_j(x)U_j(t) \tag{5.35}$$

where repeated indices indicate a summation over all possible values (summing on repeated suffixes–j in this case), N_j are finite element shape functions independent of t, and $U_j(t)$ are the nodal values independent of x, forming the vector $\mathbf{U}(t)$. Now, we can substitute (5.35) into (5.33) and (5.34) then obtain

Step1 :

$$M(\mathbf{U}^{n+\frac{1}{2}} - \mathbf{U}^n) = -\frac{1}{2}\Delta t F^n \tag{5.36}$$

Step2 :

$$M(\mathbf{U}^{n+1} - \mathbf{U}^n) = -\Delta t F^{n+\frac{1}{2}} \tag{5.37}$$

where

$$M_{ij} = \int_\Omega N_i N_j d\Omega \tag{5.38}$$

and

$$F_i^n = \int_\Omega f_x^n N_i d\Omega \tag{5.39}$$

It is clear that each step in (5.36) and (5.37) is explicit in time and may be solved independently. The term M in (5.38) is called a consistent mass matrix. If the mass matrix M is diagonal then the solution of (5.38) becomes trivial. This is often done using a lumped mass matrix resulting in an uncoupled system of equations. Of the various lumping schemes which have been proposed [8,9], one of the most efficient is obtained by distributing the element mass in proportion to the diagonal terms of the consistent mass matrix. A simple way of obtaining a lumped mass matrix is by re-placing all diagonal entries of matrix M by the sum of all the terms in the corresponding row and setting all off-diagonal terms to zero. For the 8-noded quadrilateral element and the 6-noded triangle element such lumping produces negative and zero diagonal entries. Special shape functions are used to remedy this problem [10]. The use of lumped mass matrices for transient *advection* problems generally degrades the solution by introducing oscillations and phase errors [11]. Donea *et. al.* [12] suggests an iterative explicit procedure which retains the beneficial effects of the consistent mass matrix. This procedure may be applied to an equation system given by,

$$\mathbf{M}\mathbf{u} = \mathbf{f}$$

where,

$$\mathbf{u} = \mathbf{u}_{n+1} - \mathbf{u}_n$$

(n being the time level) according to the following relation,

$$\mathbf{L}\mathbf{u}^{p+1} = \mathbf{f} - (\mathbf{M} - \mathbf{L})\mathbf{u}^p$$

here, $\mathbf{M}$ and $\mathbf{L}$ are the consistent and lumped mass matrices respectively and p is the iteration index.

5.3.4　Automatic time step selection

It is advantageous in most transient problems to adapt the time step-size to the temporal gradients of solution to reduce running costs. This may be done by simply reducing or increasing the time step-size depending upon the number of iterations required for convergence in the previous step, however this is only an ad-hoc rule. Gresho et. al. [1] have used a method based on the difference of the local time discretisation error between a predictor *i.e.*, AB (Adams- Bashforth) formula and a corrector TR (Trapezoidal rule). A description of the method is given below.

The AB formula gives for an equation, $y' = f$, (where y' represents the first derivative of y with respect to time, and so on).

$$y_{n+1}^p = y_n + \frac{\Delta t_n}{2}\left[\left(2 + \frac{\Delta t_n}{\Delta t_{n-1}}\right)y_n' - \left(\frac{\Delta t_n}{\Delta t_{n-1}}\right)y_{n-1}'\right] \quad (5.40)$$

where p stands for predictor and n is the time level.

For the local discretisation error we write the exact solution at time t_{n+1} using the Taylor series expansion,

$$y(t_{n+1}) = y_n + \Delta t_n y'_n + \frac{\Delta t_n^2}{2} y''_n + \frac{\Delta t_n^3}{6} y'''_n - O(\Delta t_n)^4 \qquad (5.41)$$

where the last term represents the error due to non-inclusion of higher order Taylor series terms. Subtracting (5.41) from (5.40) and simplifying, we get,

$$y^p_{n+1} - y(t_{n+1}) = -\frac{1}{12} \left(2 + 3\frac{\Delta t_{n-1}}{\Delta t_n} \right) \Delta t_n^3 y'''_n + O(\Delta t_n)^4 \qquad (5.42)$$

We have from the trapezoidal rule,

$$\begin{aligned} y_{n+1} &= y_n + \frac{\Delta t_n}{2} (f_{n+1} + f_n) \\ &= y_n + \frac{\Delta t_n}{2} \left(y'_{n+1} + y'_n \right) \end{aligned} \qquad (5.43)$$

Subtracting (5.41) from (5.43) and simplifying, we get,

$$y_{n+1} - y(t_{n+1}) = \frac{\Delta t_n^3}{12} y'''_n + O(\Delta t_n)^4 \qquad (5.44)$$

Subtracting (5.42) from (5.44) and eliminating the unknowns *i.e.*, y'''_n and $y(t_{n+1})$, we obtain,

$$y_{n+1} - y^p_{n+1} = 3 \left(\frac{\Delta t_n^3}{12} y'''_n \right) \left(1 + \frac{\Delta t_{n-1}}{\Delta t_n} \right) + O(\Delta t_n)^4 \qquad (5.45)$$

we now define,

$$y_{n+1} - y(t_{n+1}) = d(y_{n+1})$$

and substitute its value from (5.44) in (5.45), to obtain,

$$d(y_{n+1}) = \frac{y_{n+1} - y^p_{n+1}}{3 \left(1 + \frac{\Delta t_{n-1}}{\Delta t_n} \right)} + O(\Delta t_n)^4 \qquad (5.46)$$

From (5.44), we can write for times t_{n+1} and t_{n+2},

$$\frac{d(y_{n+2})}{d(y_{n+1})} = \left(\frac{\Delta t_{n+1}}{\Delta t_n} \right)^3 \left(\frac{y'''_{n+1}}{y'''_n} \right)$$

but,

$$y_{n+1}''' = y_n''' + O(\Delta t_n)$$

therefore,

$$\frac{d(y_{n+2})}{d(y_{n+1})} = \left(\frac{\Delta t_{n+1}}{\Delta t_n}\right)^3 + \left(\frac{\Delta t_{n+1}}{\Delta t_n}\right)^3 \left(\frac{O(\Delta t_n)}{y_n'''}\right) \tag{5.47}$$

If we specify an acceptable error ϵ for $| d(y_{n+2}) |$ and neglect the higher order term in (5.47), we can solve for the next time-step Δt_{n+1}, as,

$$\Delta t_{n+1} = \Delta t_n \left(\frac{\epsilon}{| d(y_{n+1}) |}\right)^{\frac{1}{3}} \tag{5.48}$$

The norm used is as suggested by Gresho et. al. [1]

$$| d(y_{n+1}) | = \sqrt{\frac{1}{N}\left[\frac{1}{\theta_{max}^2}\sum_{i=1}^{N} d_i^2(\boldsymbol{\theta}_{n+1})\right]} \tag{5.49}$$

Where N is the total number of nodal variables and $\boldsymbol{\theta}$ are the nodal variables. Here N and $\boldsymbol{\theta}$ do not include pressure $\mathbf{P}$.

Bixler [6] suggested the use of the mid point rule instead of the trapezoidal rule as corrector while keeping the AB formula as predictor. The midpoint rule for the equation $y' = f(y, t)$ can be written as,

$$\begin{aligned} y_{n+1} &= y_n + \Delta t_n \left(f_{n+\frac{1}{2}}\right) \\ &= y_n + \Delta t_n \left(y_{n+\frac{1}{2}}'\right) \end{aligned} \tag{5.50}$$

Bixler calculates the time truncation error for the midpoint rule by subtracting the Taylor series expansion from (5.50), as,

$$y_{n+1} - y(t_{n+1}) = -\frac{\Delta t_n^3}{24}\left(y_n''' - \frac{3\,(y_n'')^2}{y_n'}\right) + O(\Delta t_n)^4 \tag{5.51}$$

In calculating the time derivative y' for use in the predictor, he suggests a formula based on the values of y at different time levels, to improve the stability of the predictor, *i.e.*,

$$y_n' = \frac{2}{\Delta t_{n-1}}(y_n - y_{n-1}) - y_{n-1}' \tag{5.52}$$

where,

$$
\begin{aligned}
y'_{n-1} &= \frac{\Delta t_{n-2}}{\Delta t_{n-1} - \Delta t_{n-2}} \left(\frac{y_n - y_{n-1}}{\Delta t_{n-1}} \right) \\
&+ \frac{\Delta t_{n-1}}{\Delta t_{n-1} - \Delta t_{n-2}} \left(\frac{y_{n-1} - y_{n-2}}{\Delta t_{n-2}} \right)
\end{aligned}
\tag{5.53}
$$

The formula for the succeeding time-step remains the same except, $d(y_{n+1})$ is redefined for the new corrector, *i.e.*,

$$
d(y_{n+1}) = \frac{\gamma}{2 + \gamma + 3\frac{\Delta t_{n-1}}{\Delta t_n}} (y_{n+1} - y^p_{n+1})
\tag{5.54}
$$

where, $0.25 \leq \gamma \leq 1.0$. with $\gamma = 1$, the original formula is obtained. This scheme helps to increase the time-step much more rapidly when steady state impends.

For both the above time integration methods the AB formula is used as predictor which requires the time derivative, $\dot{\boldsymbol{\theta}}_o$ at the first step, which can be obtained by solving (5.19) as,

$$
\mathbf{M}_o \dot{\boldsymbol{\theta}}_o = \mathbf{F}_o - \mathbf{K}_o \boldsymbol{\theta}_o
\tag{5.55}
$$

5.4 Solution of Nonlinear System of Equations

The system of equations for incompressible flow is non-linear, and therefore an iterative solution is necessary within one time step. The fully discretised system using the generalized mid-point family of methods given in (5.25), may be written as,

$$
\left[\frac{\mathbf{M}^p_{n+\alpha}}{\Delta t} + \alpha \mathbf{K}^p_{n+\alpha} \right] (\boldsymbol{\theta}^{p+1}_{n+1}) = \left[\frac{\mathbf{M}^p_{n+\alpha}}{\Delta t} - (1 - \alpha)\mathbf{K}^p_{n+\alpha} \right] (\boldsymbol{\theta}_n) + \mathbf{F}^p_{n+\alpha}
\tag{5.56}
$$

where, p represents the iteration number. (5.56) may be solved until the norm $| \boldsymbol{\theta}^{p+1}_{n+1} - \boldsymbol{\theta}^p_{n+1} |$ falls below an acceptable tolerance.

To speed up convergence within one time-step the Newton Raphson method may be used. For an equation $y = f(x)$ the Newton Raphson method gives,

$$
f'(x^p)(x^{p+1} - x^p) = -f(x^p)
\tag{5.57}
$$

where p represents the iteration number. Applying this to a system of equations given by,

$$A(x)x = b(x)$$

we obtain,

$$\frac{\partial}{\partial x}[A(x^p)x^p - b(x)^p]\left[x^{p+1} - x^p\right] = [b(x^p) - A(x^p)x^p] \qquad (5.58)$$

The matrix corresponding to $f'(x^p)$ in (5.58) is referred to as the Jacobian matrix, **J**.

The construction of **J** for the system of equations (5.19) may be done after converting into the temporally discretised form given by (5.56). The Jacobian thus obtained is shown below.

$$\begin{bmatrix} \left(\frac{\rho \mathbf{M}_u}{\alpha \Delta t_n} + \mathbf{K}_{uu}\right) & C_u & \mathbf{K}_{uv} \\ + \int \rho N_i N_k \frac{\partial N_j}{\partial x}(u_{n+\alpha})_j & & + \int \rho N_i N_k \frac{\partial N_j}{\partial y}(u_{n+\alpha})_j \\ \\ C_u^T & 0 & C_v^T \\ \\ \mathbf{K}_{vu} & C_v & \left(\frac{\rho \mathbf{M}_v}{\alpha \Delta t_n} + \mathbf{K}_{vv}\right) \\ + \int \rho N_i N_k \frac{\partial N_j}{\partial x}(v_{n+\alpha})_j & & + \int \rho N_i N_k \frac{\partial N_j}{\partial y}(v_{n+\alpha})_j \end{bmatrix} \qquad (5.59)$$

If the automatic time selection scheme as discussed in the previous section is used, then only one iteration may be sufficient for each time-step provided ϵ is kept small. Gresho et. al. [1] suggest a value of 0.001 and refer to this method as 'One-step Newton'.

The unsymmetrical discretised system of equations obtained may be solved using direct solvers such as the frontal solution subroutine given in [13]. For large problems however iterative solvers are more suitable, as computational effort for iterative solvers increases approximately linearly (against nearly quadratically for direct solvers).

Bibliography

[1] P.M.Gresho, R.L.Lee, and R.L.Sani. On the time-dependent solution of the incompressible Navier-Stokes equations in two and three dimensions. In *Recent Advances in Numerical Methods in Fluids*, Pineridge Press Limited, Swansea, 1980.

[2] W.L.Wood. *Practical time stepping schemes*. Clarendon Press, Oxford, 1990.

[3] T.J.R.Hughes. *The Finite Element Method - Linear Static and Dynamic Finite Element Analysis*. Prentice-Hall International, Inc., Englewood Cliffs, New Jersey 07632, 1987.

[4] T.J.R.Hughes. Analysis of transient algorithms with particular reference to stability behaviour. In *Computaional methods for transient analysis*, Elsevier Science Publishers, 1983.

[5] G.Dahlquist and B.Lindberg. *On some implicit one-step methods for stiff differential equations*. Technical Report TRITA-NA-7302, Department of Information Processing, The Royal Institute of Technology, Stockholm, 1973.

[6] N.E.Bixler. An improved time integrator for finite element analysis. *Commnunications in Applied Numerical Methods*, 5:69–78, 1989.

[7] K.A.Cliffe. On conservative finite element formulations of the inviscid boussinesq equations. *International Journal for Numerical Methods in Fluids*, 1:117–127, 1981.

[8] H.C. Huang. *Static and Dynamic Analyses of Plates and Shells*. Springer-Verlag, London, 1989.

[9] O.C.Zienkiewicz E. Hinton, TA.Rock. A note on mass lumping and related processes in finite element method. *International Journal of Earthquake Engng Struct Dyn*, 4:245–249, 1976.

[10] J.Donea, S.Giuliani, and H.Laval. Accurate explicit finite element schemes for convective-conductive heat transfer problems. In T.J.R.Hughes, editor, *Finite Element Methods for Convection Dominated Flows*, ASME, AMD, 1979.

[11] P.M. Gresho and R.L.Lee. Don't suppress the wiggles - they are telling you something! *Computers and Fluids*, 9:223–253, 1981.

[12] J.Donea, S.Giuliani, H.Laval, and L.Quartapelle. Time-accurate solution of advection-diffusion problems by finite elements. *Computer Methods in Applied Mechanics and Engineering*, 45:123–145, 1984.

[13] C.Taylor and T.G.Hughes. *Finite Element Programming of the Navier-Stokes Equations*. Pineridge Press, Swansea, U.K., 1981.

Chapter 6

Finite Element Analysis for Transient Non-Newtonian Flow

6.1 Introduction

In this chapter a finite element method for analysing transient non-Newtonian flow is presented. Examples of transient non-Newtonian flows occur widely in industry in many situations *e.g.* extrusion, spinning, injection and blow moulding. Numerical simulation of such problems is facilitated through the theory embodied in Computational Fluid Dynamics. This chapter provides some insight into how this may be achieved for the generalised non-Newtonian flows by employing finite element methods in a transient solution process. The method chosen here is a Taylor-Galerkin Pressure correction method [1] which belongs to a large class of methods using *uncoupled* or *segregated* approaches (where velocity and pressure variables are uncoupled, as opposed to the mixed methods discussed earlier). The basic idea for segregated solutions stems from the work of Chorin [2]. Two main variants of this method have matured simultaneously in finite element circles over the past decade. The first method preferred by Gresho *et.al.* [3] has come to be known as the *fractional step method*, an appellation given by Donea *et.al.* [4]. The second method is sometimes referred to (as done here) as the *velocity correction method* according to Kawahara and Ohmiya [5] (the Taylor-Galerkin pressure correction method used here is a variant of this). The essential difference between the two methods is that for the former the segregation (of velocity and pressure) is effected after the GFEM discretization of the differential equations, while for the latter, it happens at the differential equation stage. The segregated methods have also been identified as *projection methods* [2,6]. This can be seen by writing the Navier Stokes equations (6.4) and (6.5) as,

$$\frac{\partial \mathbf{v}}{\partial t} + \nabla P = f(\mathbf{v}) \tag{6.1}$$

and

$$\nabla \cdot \mathbf{v} = 0 \tag{6.2}$$

where the (known) vector $f(\mathbf{v})$ which is neither divergence free ($\nabla \cdot f(\mathbf{v}) \neq 0$) nor curl free ($\nabla \times f(\mathbf{v}) \neq 0$) is decomposed (projected) onto the sum of divergence free vector ($\frac{\partial \mathbf{v}}{\partial t}$) and a curl free vector (∇P). Further discussion of this may be found in [6], in the context of the *fractional step* approach.

To illustrate the difference between the two segregated approaches mentioned above we begin with the Navier Stokes equations in the stress divergence form [7];

$$\rho \left(\frac{\partial \mathbf{v}}{\partial t} + \mathbf{v} \cdot \nabla \mathbf{v} \right) = \nabla \cdot \tau \tag{6.3}$$

which may be expanded as, with $P = \frac{p}{\rho}$,

$$\left(\frac{\partial \mathbf{v}}{\partial t} + \mathbf{v} \cdot \nabla \mathbf{v} \right) = -\nabla P + \nabla \cdot \nu \left(\nabla \mathbf{v} + (\nabla \mathbf{v})^T \right) \tag{6.4}$$

subject to the continuity (divergence free) constraint,

$$\nabla \cdot \mathbf{v} = 0 \tag{6.5}$$

We consider the velocity correction method first, and apply the forward Euler time differencing to equation (6.4) after ignoring the pressure term (∇P):

$$\frac{\mathbf{v}^* - \mathbf{v}_n}{\Delta t} = -(\mathbf{v}_n \cdot \nabla)\mathbf{v}_n + \nabla \cdot \nu \left(\nabla \mathbf{v}_n + (\nabla \mathbf{v}_n)^T \right) \tag{6.6}$$

which produces an intermediate velocity $\mathbf{v}^*$. The pressure and continuity part of the equation is now written in discrete form:

$$\frac{\mathbf{v}_{n+1} - \mathbf{v}^*}{\Delta t} = -\boldsymbol{\nabla} P_n \tag{6.7}$$

and

$$\boldsymbol{\nabla} \cdot \mathbf{v}_{n+1} = 0 \tag{6.8}$$

Taking the divergence of both sides of equation (6.7) and using equation (6.8) we obtain,

$$\nabla^2 P_n = \frac{1}{\Delta t} \boldsymbol{\nabla} \cdot \mathbf{v}^* \tag{6.9}$$

From these equations a clear procedure for computing the end of step velocities $\mathbf{v}_{n+1}$ is evident, *i.e.*

1. Explicit calculation of the intermediate velocity $\mathbf{v}^*$ via equation (6.6).

2. Pressure P_n from equation (6.9).

3. New velocities $\mathbf{v}_{n+1}$ from equation (6.7).

We have not mentioned any spatial discretization as yet for the above method and the segregation of velocity and pressure was achieved at the differential equation stage. If now the standard GFEM procedure is applied to the differential equations above we may write the solution procedure in matrix form as follows;

1. Explicit calculation of the intermediate velocity $\mathbf{v}^*$ from

$$\mathbf{M}\left(\frac{\mathbf{v}^* - \mathbf{v}_n}{\Delta t}\right) = (\mathbf{N}(\mathbf{v}_n) + \mathbf{K})\,\mathbf{v}_n \tag{6.10}$$

where $\mathbf{M}$, $\mathbf{N}$ and $\mathbf{K}$ are mass, advection and viscous diffusion matrices respectively.

2. Pressure P_n from

$$\mathbf{A}P_n = \frac{1}{\Delta t}\mathbf{C}^T \mathbf{v}^* \tag{6.11}$$

where $\mathbf{A}$ is the GFEM discretization of the Laplacian ∇^2, and $\mathbf{C}^T$ is the divergence matrix.

3. New velocities $\mathbf{v}_{n+1}$ from

$$\mathbf{M}\left(\frac{\mathbf{v}_{n+1} - \mathbf{v}^*}{\Delta t}\right) = \mathbf{C}P_n \qquad (6.12)$$

where $\mathbf{C}$ is the gradient matrix.

An equivalent version of the fractional step method in which segregation is done after spatial discretization is derived below. This approach is similar to the explicit method of Donea *et.al.* [4] and the *Projection 1* method of Gresho [6]. The spatially discretised (GFEM) form of the Navier Stokes equations (6.4) and (6.5), may be written as,

$$\mathbf{M}\dot{\mathbf{v}} + (\mathbf{N}(\mathbf{v}) + \mathbf{K})\,\mathbf{v} + \mathbf{C}P = 0 \qquad (6.13)$$

$$\mathbf{C}^T\mathbf{v} = 0 \qquad (6.14)$$

Discretising equation (6.13) in time, and ignoring the pressure term we have,

$$\mathbf{M}\left(\frac{\mathbf{v}^* - \mathbf{v}_n}{\Delta t}\right) = -\,(\mathbf{N}(\mathbf{v}_n) + \mathbf{K})\,\mathbf{v}_n \qquad (6.15)$$

which again produces an intermediate velocity $\mathbf{v}^*$. Writing the pressure and continuity parts in discrete form:

$$\mathbf{M}\left(\frac{\mathbf{v}_{n+1} - \mathbf{v}^*}{\Delta t}\right) = -\mathbf{C}P_n \qquad (6.16)$$

$$\mathbf{C}^T\mathbf{v}_{n+1} = 0 \qquad (6.17)$$

Premultiplying (6.16) by $\mathbf{C}^T\mathbf{M}^{-1}$, using (6.17) and rearranging yields,

$$\mathbf{C}^T\mathbf{M}^{-1}\mathbf{C}P_n = \frac{1}{\Delta t}\mathbf{C}^T\mathbf{v}^* \qquad (6.18)$$

which would be identical to equation (6.11) if $\mathbf{C}^T\mathbf{M}^{-1}\mathbf{C} = \mathbf{A}$. Therefore, the solution procedure for this method remains identical to the velocity correction method described earlier. Gresho [6] describes $\mathbf{C}^T\mathbf{M}^{-1}\mathbf{C}$ as the discrete (and consistent) approximation to the Laplacian ∇^2. This matrix is a truly global matrix and therefore must be constructed by multiplying the global (post assembly) versions of $\mathbf{M}^{-1}$ and $\mathbf{C}$. This makes it prohibitively expensive to use the consistent mass matrix $\mathbf{M}$ and almost invariably a lumped version is used. In spite of this, the construction of this discrete Laplacian is tedious and expensive. This is the first of several reasons in favour of the velocity correction method.

Gresho [8] has presented a detailed comparative analysis of the two methods. From this analysis, substantiated by our own experience, it turns out that mixed interpolations are still required for the fractional step method due to the existence of pressure modes. This also precludes the use of simplex elements (triangles and tetrahedra) as for the case of integrated mixed formulations. This is not the case with the velocity correction method, as pressure modes never exist. This implies that simplex elements with equal order interpolations and C^o pressures may be used. This, in our opinion, is a great advantage when complex geometries are analysed and automatic unstructured mesh generators are used to generate meshes of simplex elements. Some of the disadvantages of the velocity correction methods pointed out by Gresho [8] include, not a strict enforcement of incompressibility and poor performance at the boundaries. The first of these does not seem to be a serious problem. As yet we have not taken into account any boundary condition effects as the purpose was mainly to illustrate the two methods. The boundary conditions will be fully addressed for the explicit velocity correction method adopted for this paper.

Other [9,10] semi-implicit finite element schemes have been suggested in which the diffusion phase is implicit. These methods increase the stability limit, and larger time-steps are possible. Ramaswamy [11] has reported that the total computer time required for obtaining a steady state solution is less in the case of the semi-implicit method. The time-step size could be greater by approximately 10 and 4 times that of an explicit scheme, for low and high Reynolds number flows respectively. For high Reynolds number problems with viscosity varying as a function of temperature or velocity, as in the case of convection and turbulent flows, the semi-implicit methods would require almost the same computational time as the explicit methods since the matrices associated with the viscous terms need to computed and inverted at each time-step.

Comini and Del Giudice [12] have introduced an implicit segregated solution algorithm. They have linearised the convection terms by using the velocity values from the previous time-step. This scheme, being implicit, does not have any stability limit on the time-step size. Since the advection terms are discretised in an implicit manner, the resulting system of simultaneous equations for the intermediate velocity field is unsymmetric. Velocity correction and pressure solution involve the inversion of a system of symmetric matrices.

As mentioned earlier, in the present work, a variant of the velocity correction method (segregation at the differential equation level) has been used to formulate and solve incompressible flows. As we are interested in getting time-accurate solutions, the Taylor-Galerkin method has been employed for discretising the advection terms.

6.2   The Taylor-Galerkin Method

The Taylor-Galerkin finite element method was introduced by Donea [13] who developed this method for different time marching schemes. Donea's work was guided by the work of Morton and Parrott [14] who showed that to each particular time stepping method corresponds a different optimal form of Petrov-Galerkin weighting function. This led Donea to discretise the pure advection equation in time first, with an improved difference approximation of the time derivative term by including higher order Taylor series terms, and then using the conventional GFEM (Bubnov-Galerkin) for spatial discretisation, thus leading to the "Taylor-Galerkin" method. This method is a finite element counterpart of the Lax-Wendroff schemes used in finite differences [15]. Lohner *et al.* [16] used an Eulerian-Lagrangian approach to arrive at the same system of discretised equations obtained by Donea and justified the use of GFEM for spatial discretisation by representing the advection equation in Lagrangian coordinates and obtaining an adjoint equation. This analysis confirmed the characteristic-based nature of the Taylor-Galerkin technique. Donea *et al.* [17] and Zienkiewicz *et al.* [18] extended their respective methods to solve advection-diffusion problems. Gresho *et al.* [19] derived a similar scheme employing a 'negative diffusion' argument, generated by forward Euler time stepping, and called their correction a 'balancing tensor diffusivity'. An advantage of Taylor-Galerkin schemes is that they do not require the specification of any upwinding parameters as in some of these aforementioned methods.

The scheme chosen in this chapter has a number of novel features that distinguish it from the above. In contrast to the schemes of Donea and coworkers, an explicit approach is maintained for nonlinear terms. An implicit treatment of the second order diffusion terms results overall in a semi-implicit scheme [20]. This choice and the viscous incompressible context requires a radical departure from the type of approach of Zienkiewicz *et.al.* [18]. A pressure-correction method is introduced following Van Kan [21], with modifications as appropriate in the finite element context. This is a second-order fractional equation based scheme, that by virtue of the finite element discretisation retains continuous differential terms in the weighted residual formulation. Inherent to such operator splitting approaches, this scheme has some three phases, first and third of mass matrix form and second of Poisson type. A nonsolenoidal field variable is introduced through intermediate staged equations. An iterative solution technique (Jacobi) is employed for the mass matrix governed equations and the computation is conducted in vector-oriented form without recourse to a system matrix. A Choleski direct method is employed to solve the weak form of the Poisson equation that emerges for the pressure difference at each time step.

The following sections present details of the chosen formulation.

6.3   The Pressure Correction Method

As discussed in the introduction, the pressure correction method decouples the velocity and pressure terms of the momentum equations and implies the consideration of a Poisson equation for the pressure at each time-step.

6.3.1   Pressure correction method with midpoint rule

Beginning with the incompressible Navier-Stokes equations:

$$\rho \left(\frac{\partial \mathbf{v}}{\partial t} + \mathbf{v} \cdot \nabla \mathbf{v} \right) = \rho \mathbf{g} + \mu \nabla^2 \mathbf{v} - \nabla P \qquad (6.19)$$

With the midpoint rule or Crank Nicolson scheme with error of $0(\Delta t)$, we can write

$$\frac{\rho}{\Delta t}(\mathbf{v}^{n+1} - \mathbf{v}^n) = \left[\mu \nabla^2 \mathbf{v} - \rho \mathbf{v} \cdot \nabla \mathbf{v} \right]^n - \nabla P^{n+1} \qquad (6.20)$$

$$\nabla \cdot \mathbf{v}^{n+1} = 0 \qquad (6.21)$$

According to the projection concept we can always find an intermediate velocity field $\mathbf{v}^*$ which may satisfy the both following equations:

$$\frac{\rho}{\Delta t}(\mathbf{v}^* - \mathbf{v}^n) = \left[\mu \nabla^2 \mathbf{v} - \rho \mathbf{v} \cdot \nabla \mathbf{v} \right]^n \qquad (6.22)$$

and

$$\frac{\rho}{\Delta t}(\mathbf{v}^{n+1} - \mathbf{v}^*) = - \nabla P^{n+1} \qquad (6.23)$$

Applying ∇ operator to the both sides of the equation (6.23) and considering the equation (6.21) we obtain

$$\frac{\rho}{\Delta t} \nabla \cdot \mathbf{v}^* = - \nabla^2 P^{n+1} \qquad (6.24)$$

It is now apparent that a three step scheme can be summarised from the above operations, that is

Step1 :

$$\frac{\rho}{\Delta t}(\mathbf{v}^* - \mathbf{v}^n) = \left[\mu \nabla^2 \mathbf{v} - \rho \mathbf{v} \cdot \nabla \mathbf{v} \right]^n \qquad (6.25)$$

Step2 :

$$\nabla^2 P^{n+1} = - \frac{\rho}{\Delta t} \nabla \cdot \mathbf{v}^* \qquad (6.26)$$

Step3

$$\frac{\rho}{\Delta t}(\mathbf{v}^{n+1} - \mathbf{v}^*) - \nabla P^{n+1} \qquad (6.27)$$

Such a method is specifically designed to deal with the incompressiblity constraint and introduces a Poisson equation for the pressure ar each time-step.

6.3.2 Pressure correction method with higher order scheme

Now, let us consider a projection method of $0(\Delta t^2)$. According to the higher order scheme described in Equation (5.30) of the previous chapter, the term $\mathbf{v}^{n+1}$ is derived from $f_v^{n+\frac{1}{2}}$, therefore, the intermediate velocity field can be worked out in two steps similar to (5.33) amd (5.34). To achieve the time accuracy up to $0(\Delta t^2)$ we have to modify (6.20) to,

$$\frac{\rho}{\Delta t}(\mathbf{v}^{n+1} - \mathbf{v}^n) = \left[\mu \nabla^2 \mathbf{v} - \rho \mathbf{v} \cdot \nabla \mathbf{v}\right]^{n+\frac{1}{2}} - \theta \nabla P^{n+1} - (1-\theta) \nabla P^n$$

(6.28)

or

$$\frac{\rho}{\Delta t}(\mathbf{v}^{n+1} - \mathbf{v}^n) = \left[\mu \nabla^2 \mathbf{v} - \rho \mathbf{v} \cdot \nabla \mathbf{v}\right]^{n+\frac{1}{2}} - \nabla P^n - \theta \nabla q^{n+1} \quad (6.29)$$

where

$$\nabla q^{n+1} = \theta(\nabla P^{n+1} - \nabla P^n)$$

It may be noted that the velocity terms in Equation (6.28) fall into the higher order scheme and the pressure terms remain 1st order. Equation (6.25) is can now be split into two fractional steps as follows:

Step1a :

$$\frac{\rho}{\Delta t}(\mathbf{v}^{n+\frac{1}{2}} - \mathbf{v}^n) = \left[\mu \nabla^2 \mathbf{v} - \rho \mathbf{v} \cdot \nabla \mathbf{v} - \nabla P\right]^n$$

(6.30)

Step1b :

$$\frac{\rho}{\Delta t}(\mathbf{v}^* - \mathbf{v}^n) = \left[\mu \nabla^2 \mathbf{v} - \rho \mathbf{v} \cdot \nabla \mathbf{v}\right]^{n+\frac{1}{2}} - \nabla P^n$$

(6.31)

followed by
Step2 :

$$\theta \nabla^2 q^{n+1} = \frac{\rho}{\Delta t} \nabla \cdot \mathbf{v}^*$$

(6.32)

Step3

$$\frac{\rho}{\Delta t}(\mathbf{v}^{n+1} - \mathbf{v}^*) = -\theta \nabla q^{n+1}$$

(6.33)

6.3.3   Stability considerations

As the velocity calculation in the schemes above is carried out explicitly, the magnitude of the time-step (Δt) has to be restricted to satisfy the stability criteria for explicit methods. This can make such models more expensive than implicit ones, however as the cost of each time step is much lower here, on balance the method should remain competitive. The actual time step size limits are dictated by one of the following criteria:

Viscous diffusion

For an explicit solution using the lumped mass matrix,

$$\Delta t \leq \frac{h^2}{\nu}$$

where h is a representative size of the element.

Advection

For an iterative explicit solution,

$$C = \frac{v\Delta t}{h} \leq \frac{1}{\sqrt{3}}$$

where, C is the Courant number, and v is a representative magnitude for the velocity of the fluid in the element. Donea *et al.* [22] have given an extensive account of the stability criterion of the various schemes used for discretising the advection equation.

Further, the mesh size h should be selected in such a way that the maximum mesh Reynolds number, $(\text{Re})_g$ is within the stability limit:

$$(\text{Re})_g = \frac{vh}{\nu} \leq 1$$

The actual value of $(\text{Re})_g$ at which instabilities begin to appear depends on the velocity gradients prevailing in the solution domain. The solution will remain stable for large values of $(\text{Re})_g$ if the gradients are low.

6.4 Transient Non-Newtonian Flow

6.4.1 Transient generalised Newtonian flow

Using a semi-implicit Taylor-Galerkin pressure correction scheme [20] we can obtain the following set of equations.

1a:

$$\frac{2\rho}{\Delta t}(\mathbf{v}^{n+\frac{1}{2}} - \mathbf{v}^n) = \frac{1}{2}[\boldsymbol{\nabla}\cdot\mu\,\boldsymbol{\nabla}\mathbf{v}]^{n+\frac{1}{2}} + [\frac{1}{2}\boldsymbol{\nabla}\cdot\mu\,\boldsymbol{\nabla}\mathbf{v} - \rho\mathbf{v}\cdot\boldsymbol{\nabla}\mathbf{v} - \boldsymbol{\nabla}P]^n$$

1b:

$$\frac{\rho}{\Delta t}(\mathbf{v}^* - \mathbf{v}^n) = \frac{1}{2}[\boldsymbol{\nabla}\cdot\mu\,\boldsymbol{\nabla}\mathbf{v}^*] + [\frac{1}{2}\boldsymbol{\nabla}\cdot\mu\,\boldsymbol{\nabla}\mathbf{v} - \boldsymbol{\nabla}P]^n - [\rho\mathbf{v}\cdot\boldsymbol{\nabla}\mathbf{v}]^{n+\frac{1}{2}}$$

$$(6.34)$$

2:

$$\theta\,\boldsymbol{\nabla}^2 q^{n+1} = \frac{\rho}{\Delta t}\boldsymbol{\nabla}\cdot\mathbf{v}^* \tag{6.35}$$

3:

$$\frac{\rho}{\Delta t}(\mathbf{v}^{n+1} - \mathbf{v}^*) = -\theta\,\boldsymbol{\nabla}q^{n+1} \tag{6.36}$$

where $q^{n+1} = p^{n+1} - p^n$ and $\theta = 1/2$, $\mathbf{v}^{n+\frac{1}{2}}$ is a half step velocity field and $\mathbf{v}^*$ is an auxiliary nonsolenoidal field variable.

Now we can consider the spatial discretisation. Following the standard GFEM approach, we introduce approximations $\mathbf{U}(\mathbf{x},t)$, $\mathbf{P}(\mathbf{x},t)$ to the velocity and pressure fields respectively over finite dimensional function spaces, that is,

$$\mathbf{U}(\mathbf{x},t) = \mathbf{U}^j(t)\phi_j(\mathbf{x})$$

$$\mathbf{P}(\mathbf{x},t) = \mathbf{P}^j(t)\psi_j(\mathbf{x}) \tag{6.37}$$

where $\mathbf{U}^j(t)$, $\mathbf{P}^j(t)$ are nodal values of velocity and pressure and ϕ, ψ are their respective shape functions (same forms apply to $\mathbf{v}^*$ and q). For six noded triangular elements ϕ_j are selected as quadratic functions, whereas ψ_j are linear functions defined on vertex nodes. Substituting (6.37) into (6.34) - (6.36) and adopting the GFEM approach, a set of discretised equations is readily obtained.

6.4.2 Oldroyd's liquid B model

Among non-Newtonian fluid models, Oldroyd's B model has been discussed in Section 2.3, its constitutive equations take the form

$$\mathbf{T} + \lambda_1 \overset{\triangledown}{\mathbf{T}} = 2\mu_0 \mathbf{d} + \lambda_2 \overset{\triangledown}{\mathbf{d}} \tag{6.38}$$

where λ_1, λ_2, are material constants and μ_0 is constant viscosity, $\mathbf{T}$ is the extra-stress tensor, $\mathbf{d}$ the rate of deformation tensor and $\triangledown$ denotes the upper-convected time derivative.

Equation (6.38) can be decomposed into the following form

$$\mathbf{T} = \boldsymbol{\tau} + 2\mu_2 \mathbf{d} \tag{6.39}$$

$$\boldsymbol{\tau} + \lambda_1 \, \boldsymbol{\tau} = 2\mu_1 \mathbf{d} \tag{6.40}$$

it can be considered as a decomposition into a Newtonian solvent contribution of viscosity μ_2 together with an elastic contribution of viscosity μ_1. if we take the time differential of Equations (6.39) and (6.40) and eliminate $\boldsymbol{\tau}$ we obtain

$$\mu = \mu_1 + \mu_2 \tag{6.41}$$

$$\lambda_2 = \frac{\lambda_1 \mu_2}{\mu_1 + \mu 2} \tag{6.42}$$

Now using $\boldsymbol{\tau}$ as the unknown stress tensor we obtain the following momentum equation:

$$\rho \frac{D\mathbf{v}}{Dt} = \boldsymbol{\nabla} \cdot (\boldsymbol{\tau} + 2\mu_2 \mathbf{d}) - \boldsymbol{\nabla} P \tag{6.43}$$

which of course is coupled with the continuity equation.

The Taylor-Galerkin scheme with the projection method mentioned in the previous sections can again be used to solve the above equations.

6.5 Non-isothermal Flow

For incompressible flow under non-isothermal conditions, the energy conservation equation must also be included in the formulation, *i.e.*,

$$\rho c_p \frac{\partial T}{\partial t} = \boldsymbol{\nabla} \cdot k \, \boldsymbol{\nabla} T - \rho c_p \mathbf{v} \cdot \boldsymbol{\nabla} T + \mu \Phi \tag{6.44}$$

where T is the temperature, k is the thermal conductivity and Φ is a dissipation function dependent on shear rate as follows

$$\Phi = \frac{1}{2}I_2$$

where I_2 has been defined in (2.13).

In the non-isothermal case the Taylor-Galerkin procedure can be rewritten as

1a:
$$\frac{2\rho}{\Delta t}(\mathbf{v}^{n+\frac{1}{2}} - \mathbf{v}^n) = [\nabla \cdot \mu \nabla \mathbf{v} - \rho \mathbf{v} \cdot \nabla \mathbf{v} - \nabla P]^n$$

$$\frac{2\rho c_p}{\Delta t}(\mathbf{T}^{n+\frac{1}{2}} - \mathbf{T}^n) = [\nabla \cdot k \nabla \mathbf{T} - \rho c_p \mathbf{v} \cdot \nabla \mathbf{T} + \mu \Phi]^n$$

1b:
$$\frac{\rho}{\Delta t}(\mathbf{v}^* - \mathbf{v}^n) = [\nabla \cdot \mu \nabla \mathbf{v} - \rho \mathbf{v} \cdot \nabla \mathbf{v}]^{n+\frac{1}{2}} - \nabla P^n$$

$$\frac{\rho c_p}{\Delta t}(\mathbf{T} - \mathbf{T}^n) = [\nabla \cdot k \nabla \mathbf{T} - \rho c_p \mathbf{v} \cdot \nabla \mathbf{T} + \mu \Phi]^{n+\frac{1}{2}} \qquad (6.45)$$

2:
$$\theta \nabla^2 q^{n+1} = \frac{\rho}{\Delta t} \nabla \cdot \mathbf{v}^* \qquad (6.46)$$

3:
$$\frac{\rho}{\Delta t}(\mathbf{v}^{n+1} - \mathbf{v}^*) = -\theta \nabla q^{n+1} \qquad (6.47)$$

Identical spatial interpolation is employed for the temperature field as for velocity components,

$$\mathbf{T}(\mathbf{x}, t) = \mathbf{T}^j(t)\phi_j(\mathbf{x}). \qquad (6.48)$$

For simplicity, the preferred choice is to work in a non-dimensional frame of reference. Non-dimensionalised variables and scales are defined as follows:

$$\mathbf{x}^* = \frac{\mathbf{x}}{L}, \qquad \mathbf{v}^* = \frac{\mathbf{v}}{V}, \qquad t^* = \frac{tV}{L}, \qquad p^* = \frac{P}{\rho V^2},$$
$$T^* = \beta(T - T_0),$$

where L is a characteristic length, V is a characteristic velocity, T_0 is a reference temperature and β is a temperature coefficient of viscosity. Accordingly it is appropriate to define a characteristic shear rate as

$$\bar{\dot{\gamma}} = \frac{V}{L}$$

and a characteristic viscosity as

$$\bar{\mu} = \mu(\bar{\dot{\gamma}}, T_0).$$

Therefore, for a Power-Law fluid with Arrhenius Law

$$\bar{\mu} = \mu_0 \bar{\dot{\gamma}}^{m-1} e^{-\beta(T-T_0)} = \mu_0 \left(\frac{V}{L}\right)^{m-1} e^{-\beta(T-T_0)}$$

where μ_0 is a constant viscosity reference value associated with a unit shear rate and the reference temperature T_0[23].

Three nondimensional groups may be identified. The Reynolds number may be defined by

$$\text{Re} = \frac{\rho V L}{\bar{\mu}},$$

the thermal Peclet number as

$$\text{Pe} = \frac{c_p^*}{k^*} = \frac{\rho c_p L V}{k}$$

and the Brinkman number as

$$\text{Br} = \frac{\bar{\mu} V^2}{k(T - T_0)}$$

in which

$$\mu^* = \frac{\mu}{\bar{\mu}},$$

$$k^* = \frac{k}{\rho V^3 L \beta},$$

$$c_p^* = \frac{c_p}{V^2 \beta}.$$

Substitutions of the above in the governing equations yield,

$$\frac{D\mathbf{v}^*}{Dt^*} = \frac{1}{\text{Re}} \nabla \cdot \mu^* \nabla \mathbf{v}^* - \nabla P^*$$

$$\frac{DT^*}{Dt^*} = \frac{1}{\text{Pe}} \nabla \cdot \nabla T^* + \frac{\text{Br}}{\text{Pe}} \Phi^* \qquad (6.49)$$

6.6 Adaptive Analysis

Techniques for error estimation have recently been developed for stress analysis [24], heat transfer [25] and fluid flow problems [26]. These methods may be considered as optimizing the finite element analysis according to the intrinsic behaviour of the given problem. The type of error estimation with adaptive remeshing is now applied to non-isothermal Generalised Newtonian fluids where the accuracy of both velocity and temperature is important. The error norm is defined by gradients of velocity and temperature, therefore, The technique will calculate errors in velocity and temperature and predict the corresponding element sizes to generate an adaptive mesh.

The size and placement of the elements largely determines the accuracy with which a problem can be solved. Reducing the element size and thereby increasing the number of nodal points usually yields a more accurate solution but at the cost of an increased CPU time and memory requirement. The key to the efficient and economic solution of problems is not merely the *number* of nodal points and elements but also their *placement*. Regions with large gradients (e.g. a discontinuity, stress concentration or region of high heat flux) will need a high mesh density, with quiescent regions requiring a comparatively coarser mesh. In many real engineering situations, attempts will be made either to obtain the most accurate solution possible within an upper problem size limit, or to try to capture a numerically awkward feature without using an excessive number of elements.

The premise of adaptive procedures is that, by making use of the mathematics of error analysis, a finite element program can automatically adapt the mesh to suit the problem. The error is calculated in each element and is compared with a predefined limit. It is expected that for a given accuracy the *optimal* mesh will contain approximately the same level of error in each element. Therefore, any element with an error above, or below, must be adjusted to match this error level. The process is repeated, if necessary, with the ultimate aim that every element contains the same predefined, allowable error, thus yielding an optimal mesh.

The adaptive strategy employed generally follows the well established procedures developed by Zienkiewicz and Zhu [27]. Several distinct phases of the adaptive strategy can be described as an iterative process as follows:

1. **Preliminary analysis**

2. **Recovery** of the field variable gradients from the preliminary analysis.

3. *A posteriori* **error analysis** using the recovered gradients.

4. **Mesh refinement** based on the estimated error.

5. **Data transfer** to the refined mesh (for nonlinear/transient problems).

6. **Reanalysis** and back to step 2 if convergence not achieved.

Step 1 is essentially the same as a standard FE analysis. Theoretically it is unnecessary at this stage to worry about the element size distribution in the mesh as the adaptive procedure will automatically converge to the *optimal mesh* (where each element has approximately the *same* error). However in practice it is beneficial even at this stage to create the mesh with due care and attention to possible areas of high gradients as this will help reduce the number of cycles required for convergence to the optimal mesh. The subsequent steps involved in the strategy are discussed in the following paragraphs.

In the absence of the *exact* solution a higher order projection of the available FE solution is used. This process is termed as *smoothing* or *recovery*. The global smoothing technique of Hinton and Campell [28] was the first, and has remained popular. A variety of gradient recovery procedures have since appeared in the FE literature which are either better [29] or more efficient or in some cases both [30]. It has been proved that the value of the field variable obtained by the finite element method is most accurate at nodal points [31], whereas its gradient is most accurate at Gauss points [32]. This is often referred to as the superconvergence phenomenon. The Gauss points have been shown to be *super-convergent* points. Based on this Zienkiewicz and Zhu introduced a local recovery procedure called *super-convergent patch recovery* [30]. In this method *smooth* nodal gradients are generated by a least square fit from super-convergent gradient values at Gauss points on a patch of elements surrounding each node. This method of recovery is more efficient and allows faster convergence towards the optimal mesh and has been used for this work.

Error analysis allows us to calculate for each element a value of error relative to other elements in the mesh. Much of the early mathematical work in error analysis was due to Babuska and Rheinboldt [33,34], but was drawn together, in the form of an error estimate, in a notable paper by Kelly *et.al.* [35]. A companion paper by Gago *et.al.* [36] was later published which offered several strategies for using an error estimator to refine a mesh.

Once an estimate of the error in each element has been obtained, there are several ways of going about the next phase of the adaptive process, i.e. improving the mesh. The two basic approaches are *p-refinement* and *h-refinement*. *p*-refinement involves increasing the order of the polynomial approximation within an element while *h*-refinement simply means reducing the subdivision size. *p*−refinement, especially when combined with a hierarchical formulation, has several advantages; it is more efficient, converges faster and appeals to the purist by virtue of its mathematical elegance. However, incorporating *p*−refinement generally means restructuring, if not rewriting, an existing finite element code. *h*−refinement is more universally accepted and has been employed in the content of this chapter.

Having decided on adopting the *h*−refinement method there is another choice to be made, *element subdivision* or *mesh regeneration*. With the element subdivision (*mesh enrichment*) approach, every element that exceeds the allowable error threshold is subdivided into smaller elements. This is most effective when using four-node elements as it is otherwise very difficult to achieve the desired density distribution. However, constrained nodes are introduced which must be dealt with, and the method allows only one level of subdivision at a time. For large problems it is also more efficient to refine by *mesh movement* and/or *mesh enrichment* however there are inherent limitations to these approaches.

The remeshing approach involves completely regenerating the mesh, either in regions of high error only, or over the entire domain.

This can be expensive, however the advantage of regenerating the entire mesh is that areas can be coarsened if the calculated error is below the allowable error. This allows the generation of a truly optimal mesh in which every element has approximately the same, predefined, level of error. Another disadvantage of mesh regeneration is that a highly robust mesh unstructured generators capable of creating meshes of sharply varying element sizes within relatively small distances are a necessity to allow full use of the abundant information provided by an error estimation procedure. This is only possible with simplex elements (triangles and tetrahedra). A number of automatic mesh generation algorithms are available such as the Delaunay triagulation based automatic mesh generator used in another book by the authors [37] and the advancing front method [38,39].

For transient or non-linear problems one needs to interpolate the nodal information from the previous mesh to the new mesh. The interpolation for linear triangular elements may be conducted as follows

$$\phi(x,y) = L_1(x,y)\phi_1 + L_2(x,y)\phi_2 + L_3(x,y)\phi_3 \qquad (6.50)$$

where L_1, L_2, L_3 are area coordinates at (x,y) and ϕ_1, ϕ_2, ϕ_3 are values at the element nodes. Such linear interpolation is not sufficient for many problems. A higher order interpolation for linear triangular elements may be written as,

$$\phi(x,y) = \sum_{i=1}^{3} N_i(x,y)\left(\phi_i + R_i(x,y)\right) \qquad (6.51)$$

where N_i are the same as L_i for the linear triangular element, and R_i are functions of the gradients of ϕ [37],

$$R_i(x,y) = 0.5\left((x - x_i)\left(\frac{\partial\phi}{\partial x}\right)_i + (y - y_i)\left(\frac{\partial\phi}{\partial y}\right)_i\right) \qquad (6.52)$$

here the gradients of ϕ do not have to be calculated as they are available from the preceding error estimation process.

For the problems considered in this text adaptivity has only been used after reaching converged steady state solutions and therefore

it has not been necessary to use this feature. If however, one needs
to obtain a high resolution spatial solution evolving in time, quantities will need to be transferred from one mesh to another. This
can lead to convergence problems in highly non-linear problems and
may require data transfer schemes that conserve a greater number
of the Fourier components of the field to be transferred (relative
to the methods above).

The strategy described above is now applied to generalised Newtonian flow. The details of this work is presented as follows.

6.7 Error Estimation for Fluid Flow

The error estimation procedure for scalar diffusion problems (heat
transfer) developed by Huang and Lewis [25] was extended to fluid
flow in [26]. Following this work the error norm in a two dimensional element Ω_e is defined as

$$
\begin{aligned}
||e||^2_e \ = \ \mu \int_{\Omega_e} [(\nabla u - \nabla \hat{u})^T (\nabla u - \nabla \hat{u}) \\
+ (\nabla v - \nabla \hat{v})^T (\nabla v - \nabla \hat{v})] d\Omega_e
\end{aligned}
\tag{6.53}
$$

where u and v represent the 'true' velocity components and $\hat{u}, \hat{v}$
the solution obtained from the finite element analysis. The exact
values of the gradients $\nabla u, \nabla v$ must of course be approximated by
smoothed values [24]. If we define a quantity q in each element as
follows,

$$
||q||^2_e = \mu \int_{\Omega_e} \left[(\nabla \hat{u})^T \nabla \hat{u} + (\nabla \hat{v})^T \nabla \hat{v} \right] d\Omega_e
\tag{6.54}
$$

the error can be described in terms of a percentage error (η),

$$
\eta = \frac{\sum ||e||}{\sum ||q||} \text{x} 100\%
\tag{6.55}
$$

where the $\sum$ is over all elements. As gradients of the velocities are
used here, we can call this a *gradient scheme*.

Alternatively, Wu *et.al.* [40] have suggested the use of an 'energy
norm', which is simply a extention of the original Zinekiewicz-Zhu

error estimation [27] for solid mechanics applications. In which case, the error in the energy norm is defined as,

$$||e||^2_e = \frac{1}{2\mu} \int_{\Omega_e} \left[(\boldsymbol{\tau})^T(\boldsymbol{\tau}) \right] d\Omega_e \qquad (6.56)$$

where $\boldsymbol{\tau}$ is the vector of extra stresses. The energy norm in each element is written as

$$||q||^2_e = \frac{1}{2\mu} \int_{\Omega_e} \left[(\boldsymbol{\tau} - \hat{\boldsymbol{\tau}})^T(\boldsymbol{\tau} - \hat{\boldsymbol{\tau}}) + p\nabla \cdot \mathbf{v} \right] d\Omega_e \qquad (6.57)$$

We can call this method a *stress scheme*.

To set a target for the adaptive analysis to achieve, a maximum permissible error $\bar{\eta}$, is specified. The requirement for a near optimum analysis is that all the elements of the final mesh must contain an approximately equal error. The maximum permissible error for each element can be calculated by distributing $\sum ||q||^2$ equally over all the elements, i.e.

$$||\bar{e}||^2_e \le \bar{\eta} \left(\frac{\sum ||q||^2}{m} \right) \qquad (6.58)$$

The error in each element is compared to the maximum permissible error in an element as calculated above and used to modify the mesh for a repeat analysis. If we define a variable ξ_e, where

$$\xi_e = \frac{||e||_e}{||\bar{e}||_e} \qquad (6.59)$$

then if $\xi_e > 1$, the size of element e must be reduced and the mesh will require refinement, otherwise, the size of the element must be increased and the mesh will be coarsened. Thus the predicted size of the new element can be calculated from the current element size as follows

$$\bar{h}_e = \frac{h_e}{\xi_e^{1/P}} \qquad (6.60)$$

where $\bar{h}_e$ is the predicted element size, h_e is current element size and P is the order of the shape functions.

In a non-isothermal fluid, sharp boundary layers may exist due to large temperature gradients. It may therefore be important to include temperature in the error estimation process for such problems. This is not straightforward using the stress scheme, however, it may be achieved quite simply with the gradient scheme as follows.

$$\|e\|^2_e = \int_{\Omega_e} [(\nabla u - \nabla \hat{u})^T \mu (\nabla u - \nabla \hat{u})$$
$$+ (\nabla v - \nabla \hat{v})^T \mu (\nabla v - \nabla \hat{v})] d\Omega_e$$
$$+ \int_{\Omega_e} \left[(\nabla T - \nabla \hat{T})^T k (\nabla T - \nabla \hat{T}) \right] d\Omega_e \quad (6.61)$$

The quantity q in each element is now written as,

$$\|q\|^2_e = \int_{\Omega_e} \left[(\nabla \hat{u})^T \mu \nabla \hat{u} + (\nabla \hat{v})^T \mu \nabla \hat{v} + (\nabla \hat{T})^T k \nabla \hat{T} \right] d\Omega_e \quad (6.62)$$

6.8 Numerical Examples

The example of flow over a step solved earlier for steady flow, is now re-examined here. Newtonian and non-Newtonian behaviour will be compared. The advantages of adaptive analysis will also presented using this example.

6.8.1 Newtonian flow

The boundary conditions remain as given in Figure 4.2 of chapter 4.3.3, however, the channel length has been extended to 16D to achieve fully developed velocity profile at exit. The element mesh is the same as in Figure 4.2 for channel length up to 4D and beyond this location an identical extension is added. Characteristic parameters as described in Section 6.4 are: L=D=1m, V=1 m/s and Re = 1, 100 and 200 for general power law flow with p=1.

The results in Figures 6.1 and 6.2 are shown only up to the length of 4D, which gives the most representative flow behaviour. The stream lines are given in Figure 6.1 representing the stream function ϕ=0.1 to 0.9 with increments of 0.1. The values of stream lines for the vortex in Figure 6.1.b are -0.003 for out side line and

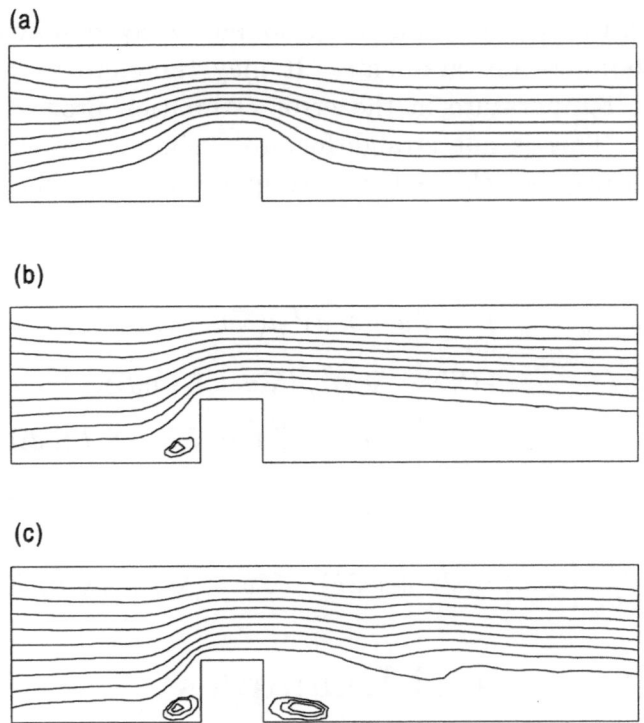

Figure 6.1: Stream lines of the flow over a step: a) Re=1; b) Re=100; c) Re=200, for Newtonian flow.

-0.006 for inside one, while for those in Figure 6.1.c they are -0.003, -0.009 and -0.015. It can be seen that the vortex disappears for lower Re (Re=1 in Figure 6.1.a). Only one vortex forms before the step for Re=100 in Figure 6.1.b, however, for Re=200 a vortex also formed after the step. Figures 6.2.a, 6.2.b and 6.2.c show the velocity profiles in the x-direction. The effect of Re on the development of the profiles is clearly seen at the 4D location (right hand side profiles in the figure).

6.8.2   Non-Newtonian flow

For non-Newtonian flow we again adopted the Power law equation with an index of 0.75. Figure 6.3 and 6.4 show the stream lines and velocity profiles with Re=1, 50 and 100. The time step was cho-

(a)

(b)

(c)

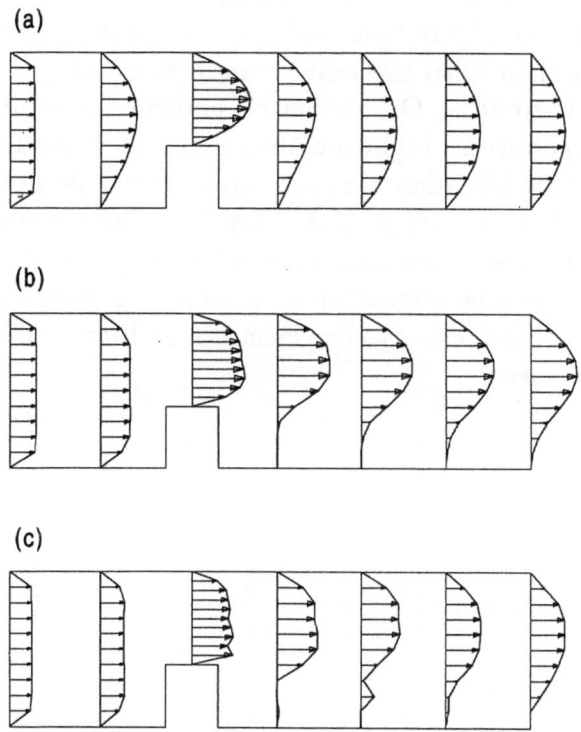

Figure 6.2: Velocity profiles of the flow over a step: a) Re=1; b) Re=100; c) Re=200, for Newtonian flow

sen to be 0.1 and solutions converged after about 4200 timesteps. There was no vortex for Re=1. A vortex appears on near side of the step for Re=50 and on both sides for Re=100.

6.8.3   Adaptive method

Adaptive analysis of the problem of Figure 4.2 was performed with a channel length of 4D. The configuration of flow over a step, though not very complicated, does not immediately suggest the type of mesh that must be used for a reasonable solution. For this particular problem the choice of a good mesh is further complicated by the non-Newtonian behaviour, that is shear thinning viscosity. An adaptive remeshing procedure is well suited to such problems. The target was set to 20% of the norm error (equivalent to 4.5

per cent error in velocity). The initial finite element mesh for this problem is shown in Figure 6.5a which gives the norm error η of 83%. The norm error with the refined mesh B is reduced to 55% and with Mesh C to 38%. On further refinement the solution approaches the target error. Figures 6.5b-d show the refined meshes, obtained from using the adaptive procedure. It may be noted that a very fine mesh is produced near the step boundaries where large velocity gradients exist. Vortices progressively appeared on both sides of the step during remeshing as shown in Figure 6.6. The development of the velocity profiles is shown in Figure 6.7 for the four adaptive meshes.

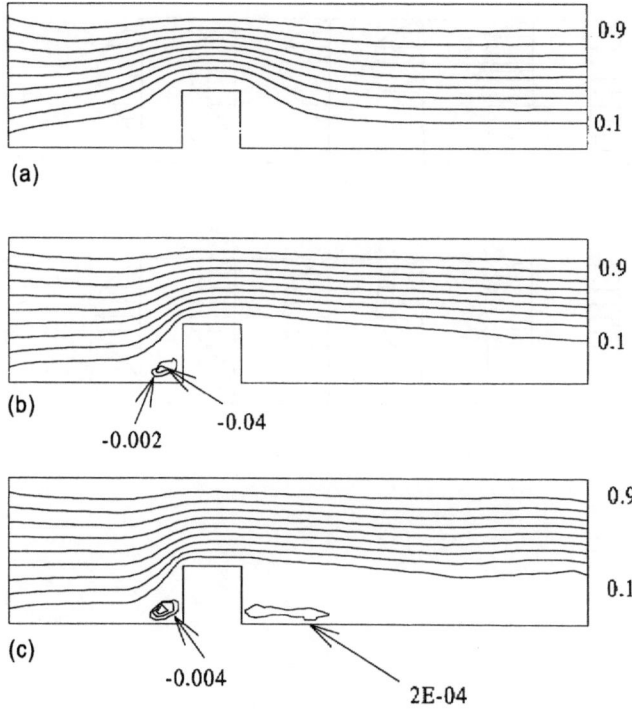

Figure 6.3: Stream lines of the flow over a step: a) Re=1; b) Re=50; c) Re=100, for power law index of 0.75

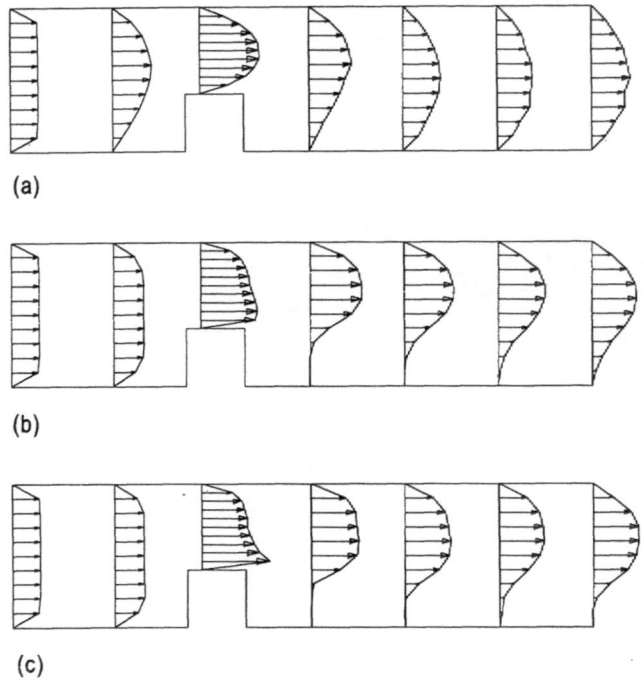

(a)

(b)

(c)

Figure 6.4: Velocity profiles of the flow over a step: a) Re=1; b) Re=50; c) Re=100, for power law index of 0.75.

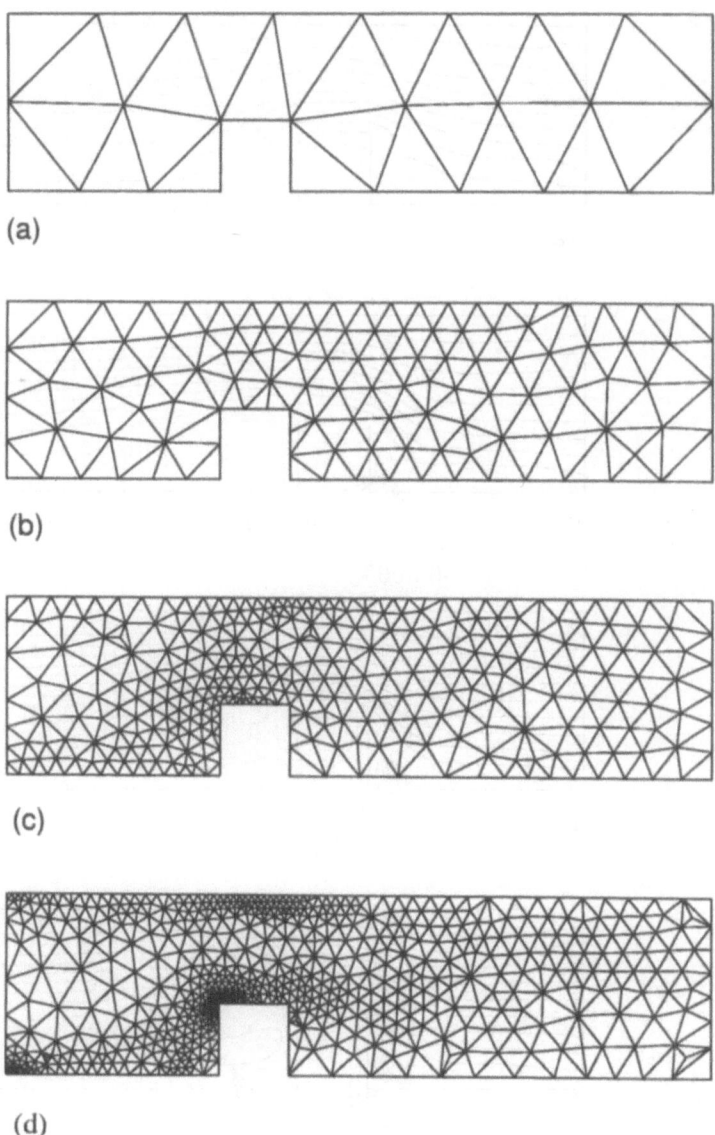

Figure 6.5: Showing the progressive refinement of the element meshes using adaptive techniques.

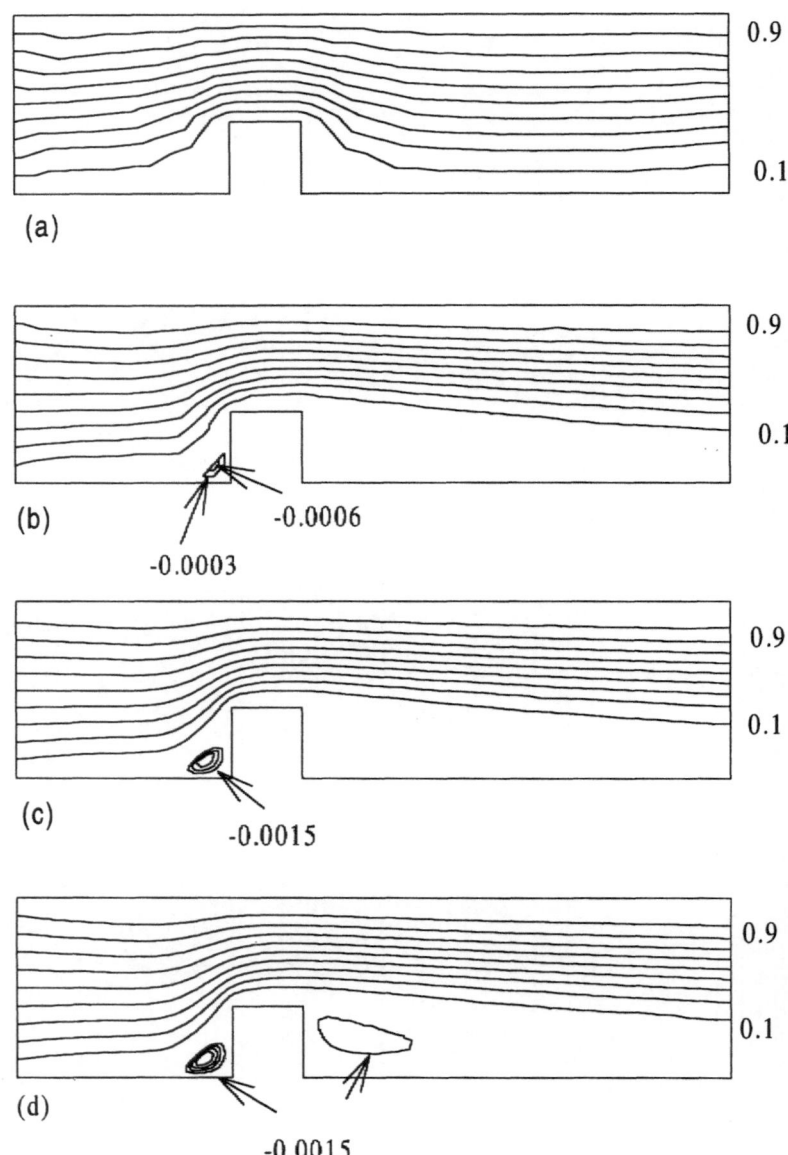

Figure 6.6: Stream lines corresponding to the four adaptive meshes

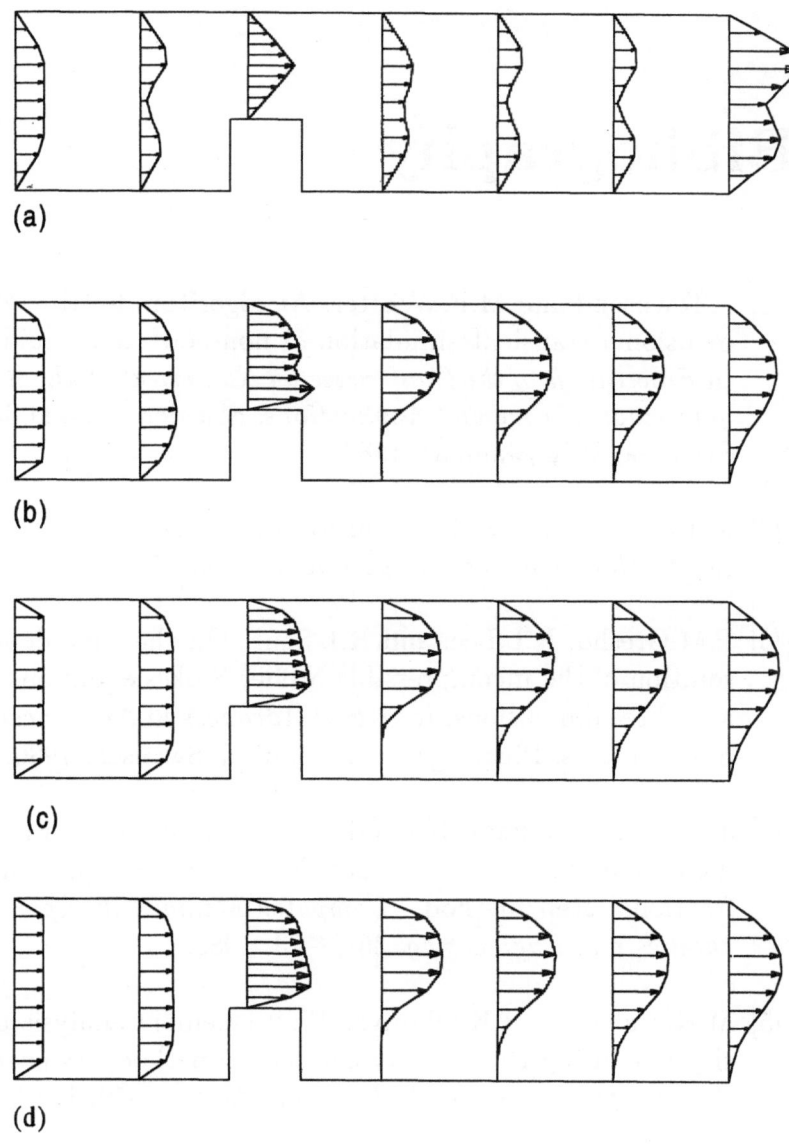

Figure 6.7: Velocity profiles corresponding to the four adaptive meshes

Bibliography

[1] P.Townsend and M.F.Webster. An algorithm for the three dimensional transient simulation of non-newtonian fluid flows. In *Proceedings of the Conference on Numerical Methods in Engineering: Theory and Applications*, Elsevier Applied Science, Swansea, U.K., January 1987.

[2] A.J.Chorin. Numerical solution of the Navier-Stokes equations. *Mathematical Computation*, 22:745–762, 1968.

[3] P.M.Gresho, R.L.Lee, and R.L.Sani. On the time-dependent solution of the incompressible Navier-Stokes equations in two and three dimensions. In *Recent Advances in Numerical Methods in Fluids*, Pineridge Press Limited, Swansea, 1980.

[4] J.Donea, S.Giuliani, H.Laval, and L.Quartapelle. Finite element solution of the unsteady Navier-Stokes equations by a fractional step method. *Computer Methods in Applied Mechanics and Engineering*, 30:53–73, 1982.

[5] M.Kawahara and K.Ohmiya. Finite element analysis of density flow using the velocity correction method. *International Journal for Numerical Methods in Fluids*, 6:659, 1986.

[6] P.M.Gresho. On the theory of semi-implicit projection methods for viscous incompressible flow and its implementation via a finite element method that also introduces a consistent mass matrix. part 1: theory and part 2: implementation. *International Journal for Numerical Methods in Fluids*, 11:587–659, 1990.

[7] R.W.Lewis, K.Ravindran, and A.S.Usmani. Finite element solution of incompressible flows using an explicit segregated approach. *Archives of Computational Methods in Engineering*, 2:69–93, 1995.

[8] P.M.Gresho. An analysis of the velocity correction method of kawahara *et. al. Bulletin of the Faculty of Science and Engineering*, Chuo university, Tokyo, 30:43–61, 1987.

[9] B.Ramaswamy and T.C.Jue. Some recent trends and developments in finite element analysis for incompressible thermal flows. *International Journal for Numerical Methods in Engineering*, 35:671–707, 1992.

[10] P.M.Gresho. On the theory of semi-implicit projection methods for viscous incompressible flow and its implementation via a finite element method that also introduces a consistent mass matrix. part 1: theory and part 2: implementation. *International Journal for Numerical Methods in Fluids,* 11:587–659, 1990.

[11] B.Ramaswamy and T.C.Jue. Theory and implementation of a semi-implicit finite element method for viscous incompressible flow. *Computers and Fluids,* 22:725–747, 1993.

[12] G.Comini and S.Del Giudice. Finite element solution of the incompressible Navier-Stokes equations. *Numerical Heat Transfer*, 5:463–478, 1982.

[13] J.Donea. A Taylor-Galerkin method for convective transport problems. *International Journal for Numerical Methods in Engineering*, 20:101–119, 1984.

[14] K.W.Morton and A.K.Parrott. Generalized Galerkin methods for first order hyperbolic equations. *Journal of Computational Physics*, 36:249–270, 1980.

[15] P.J.Roache. *Computational Fluid Mechanics.* Hermosa Publishers, Albuquerque, U.S.A., 1976.

[16] R.Lohner, K.Morgan, and O.C.Zienkiewicz. The solution of non-linear hyperbolic equation systems by the finite element method. *International Journal for Numerical Methods in Fluids*, 4:1043–1063, 1984.

[17] J.Donea, S.Giuliani, H.Laval, and L.Quartapelle. Time-accurate solution of advection-diffusion problems by finite elements. *Computer Methods in Applied Mechanics and Engineering*, 45:123–145, 1984.

[18] O.C.Zienkiewicz, R.Lohner, K.Morgan, and J.Peraire. High-speed compressible flow and other advection dominated problems of fluid dynamics. In R.H.Gallagher, G.Carey, J.T.Oden, and O.C.Zienkiewicz, editors, *Finite Elements in Fluids*, John Wiley and Sons, 1985.

[19] P.M. Gresho, S.T.Chan, R.L.Lee, and C.D.Upson. A modified finite element method for solving the time-dependent, incompressible Navier-Stokes equations. *International Journal for Numerical Methods in Fluids*, 4:557–598, 1984.

[20] D.M.Hawken,   H.R.Tamaddon, P.Townsend, and M.F.Webster. Taylor-galerkin based algorithm for viscous incompressible flow. *International Journal for Numerical Methods in Fluids*, 10:327–351, 1990.

[21] J.Van Kan. A second order accurate pressure-correction scheme for viscous incompressible flow. *SIAM J. Sci. Stat. Comput.*, 7:870–891, 1986.

[22] J.Donea, L.Quartapelle, and V.Selmin. An analysis of time discretization in the finite element solution of hyperbolic problems. *Journal of Computational Physics*, 70:463–499, 1987.

[23] E. Mitsoulis, R. Wagner, and F.L. Heng. Numerical simulation of wire-coating low-density polyethylene:theory and experiments. *Polymer Engineering and Science*, 28:291–310, 1988.

[24] O.C.Zienkiewicz and J.Z.Zhu. A simple error estimator and adaptive procedure for practical engineering analysis. *Interna-*

tional Journal for Numerical Methods in Engineering, 24:337–357, 1987.

[25] H.C.Huang and R.W.Lewis. Adaptive analysis for heat flow problems using error estimation techniques. In *Sixth International Conference for Numerical Methods in Thermal Problems*, Pineridge Press, Swansea, Swansea, U.K., July 1989.

[26] P.Townsend H.C.Huang and M.F.Webster. Finite element simulation of complex polymer flows using structured and unstructured meshes. In *Proceeding of Numerical grid generation in Computational fluid dynamics and related fields*, Swansea, UK, April 1994.

[27] O.C.Zienkiewicz and J.Z.Zhu. A simple error estimator and adaptive procedure for practical engineering analysis. *International Journal for Numerical Methods in Engineering*, 24:337–357, 1987.

[28] E.Hinton and J.S.Campbell. Local and global smoothing of discontinuous finite element functions using a least squares method. *International Journal for Numerical Methods in Engineering*, 8:461–480, 1974.

[29] A.Tessler, H.R.Riggs, and S.C.Macy. A variational method for finite element stress recovery and error estimation. *Computer Methods in Applied Mechanics and Engineering*, 111:369–382, 1994.

[30] O.C.Zienkiewicz and J.Z.Zhu. The superconvergent patch recovery and *a posteriori* error estimates. Part 1: The recovery technique, Part 2: Error estimates and adaptivity. *International Journal for Numerical Methods in Engineering*, 33:1331–1382, 1992.

[31] T.J.R.Hughes. *The Finite Element Method - Linear Static and Dynamic Finite Element Analysis*. Prentice-Hall International, Inc., Englewood Cliffs, New Jersey 07632, 1987.

[32] M. Zlamal. Superconvergence and reduced integration in the finite element method. *Mathematics of Computation*, 32:663–685, 1978.

[33] I. Babuska and W.C. Rheinboldt. Adaptive approaches and reliability estimates in finite element analysis. *Computer Methods in Applied Mechanics and Engineering*, 17/18:519–514, 1979.

[34] I. Babuska and W.C. Rheinboldt. Error estimates for adaptive finite element computations. *SIAM Journal of Numerical Analysis*, 15, 1978.

[35] D.W. Kelly, J.P. De S.R. Gago, O.C. Zienkienwicz, and I. Babuska. A posteriori error analysis and adaptive processes in the finite element method: part 1 — error analysis. *International Journal for Numerical Methods in Engineering*, 19:1596–1619, 1983.

[36] J.P. De S.R. Gago, D.W. Kelly, O.C. Zienkienwicz, and I. Babuska. A posteriori error analysis and adaptive processes in the finite element method: part 2 — adaptive mesh refinement. *International Journal for Numerical Methods in Engineering*, 19:1621–1656, 1983.

[37] H.C.Huang and A.S.Usmani. *Finite Element Analysis for Heat Transfer - Theory and Software*. Springer-Verlag, London, 1994.

[38] J. Peraire, M. Vahdati, K. Morgan, and O.C. Zienkiewicz. Adaptive remeshing for compressible flow computations. *Journal of Computational Physics*, 72:449–466, 1987.

[39] J. Peraire, K. Morgan, and J. Piero. Unstructured finite element mesh generation and adaptive procedures for CFD. In *AGARD FDP: Specialists meeting*, Loen, Norway, May 1989.

[40] J.Wu, J.Z.Zhu, J.Szmelter, and O.C.Zienkiewicz. Error estimation and adaptivity in Navier-Stokes incompressible flows. *Computational Mechanics*, 6:259–270, 1990.

Appendix A

Software Description for NSTEAD

A.1 Introduction

Description of the program **NSTEAD.FOR** and instructions for users have been given in this appendix. The program allows a user to perform a steady flow analysis for 2-D plane or axisymmetric problems. Three and six noded triangular elements are implemented in the program. Boundary conditions permitted are fixed velocities and slip boundaries with friction.

A.2 Glossary of Variable Names

A brief description of the main integer and real variables, main integer and real arrays and the main subroutines is given in the following sections in alphabetical order.

A.2.1 Main Variables

Variable name	Description
GRAVY	Gravity constant
IEDGE	Number of edges with tractions

Variable name	Description
IGRAV	For gravity case if > 0
IPLOD	Number of point loads
MBUFA	Maximum number of buffer size
MELEM	Maximum number of elements
MEVAB	Maximum variables per element
MFRON	Maximum front width when frontal solver is used
MITER	Maximum number of iterations
MMATS	Maximum number of materials
MPOIN	Maximum number of nodes
MSTIF	Maximum space specified for global stiffness matrix
MTOTG	Maximum number of Gauss points of whole domain (7*MELEM)
MTOTV	Maximum number of unknowns (velocities)
MVFIX	Maximum number of boundary nodes
NBELM	Number of elements on slip boundaries
NCONV	1 for convection case; 0 for non-convection
NDOFN	Number of freedoms per node (2)
NEASS	Current number of element with traction loads
NEDGE	Number of edges with tractions
NELEM	Number of elements
NGAUS	Number of integration points per element
NLINR	0 for linear and 1 for non-linear problems
NMATS	Number of materials
NNODE	Number of nodes per element
NNODP	Number of pressure nodes per element
NODEG	Number of nodes per side of an element
NPOIN	Number of nodes
NPOWR	1 for power law Newtonian flow; 0 for Newtonian flow
NPROP	Number of properties for material set (8)
NSTRE	Number of stresses: plane problem using 3; axialsymmetric using 4

Variable name	Description
NTYPE	Models: 1 = plane stress only if nlinr=0; 2 = plane problem; 3 = axialsymmetric problem
NVFIX	Number of boundary nodes
PFRIC	Friction coefficient
RAMUV	Ram speed (averaged inlet speed)
THETA	Gravity angle
TOLER	Convergence tolerance

A.2.2 Main Arrays

Array name	Description
COORD(MPOIN, MDIME)	Nodal coordinates
IEFIC(100,5)	Friction element information
LNODS(MELEM,9)	Element nodal connectivities
MATNO(MELEM)	Element material type
NOFIX(MVFIX)	Node number at fixed velocity boundary
NOPRS(4)	Store nodes at the edges with tractions
NOUTP(2)	Output parameters
PRESC(MVFIX,3)	Boundary velocity values
PRESS(4,2)	Normal and tangetial distributed loads at IODEGth node
PROPS(MMATS, NPROP)	Material properties

A.2.3 Main Subroutines

Subroutine	Description
BMATPS	Evaluates the strain rate-velocity matrix
BUMTX	Deals with friction effect
CALSTR	Calculates the strain rates and the stresses and evaluates the pressures

Subroutine	Description
CHECK1	Checks the main control data
CHECK2	Checks the remainder of the input data
CONVER	Checks for convergence of the iteration process
CONVCF	Evaluates the consistent nodal forces for each element
FRONT	Undertakes equation solution by the frontal method
GAUSSQ	Sets up the Gauss-Legendre integration consts.
INDATA	Increments the applied loading
INPUT	Accepts most of the input data
INVAR	Evaluates the stress invariants and the current value of the yield function
JACOB2	Evaluates the jacobian matrix and the cartesian shape function derivatives
LINEAR	Evaluates stresses and strain rates assuming linear elastic behaviour
LIŃSH3	For boundary interpolation
LOADPS	Evaluates the consistent nodal forces for each element
NODVAL	Calculates the values at nodes from gauss poins
OUTPUT	Outputs velocities,reactions and stresses
SETUP	Sets boundary conditions
SFR2	Evaluates shape functions and their derivatives for linear, quadratic and cubic lagrangian and serendipity isoparametric 2-d elements
STIFFP	Evaluates the stiffness matrix for each element in turn
TRANSE	Transforms to slope coordinates
TRANXY	Resets velocity to x-y coordinates
ZERO	Initialises various arrays to zero

A.3 Program Overview

The main routine is listed in the following lines, the program structure with all the subroutines is shown in Figure A.1.

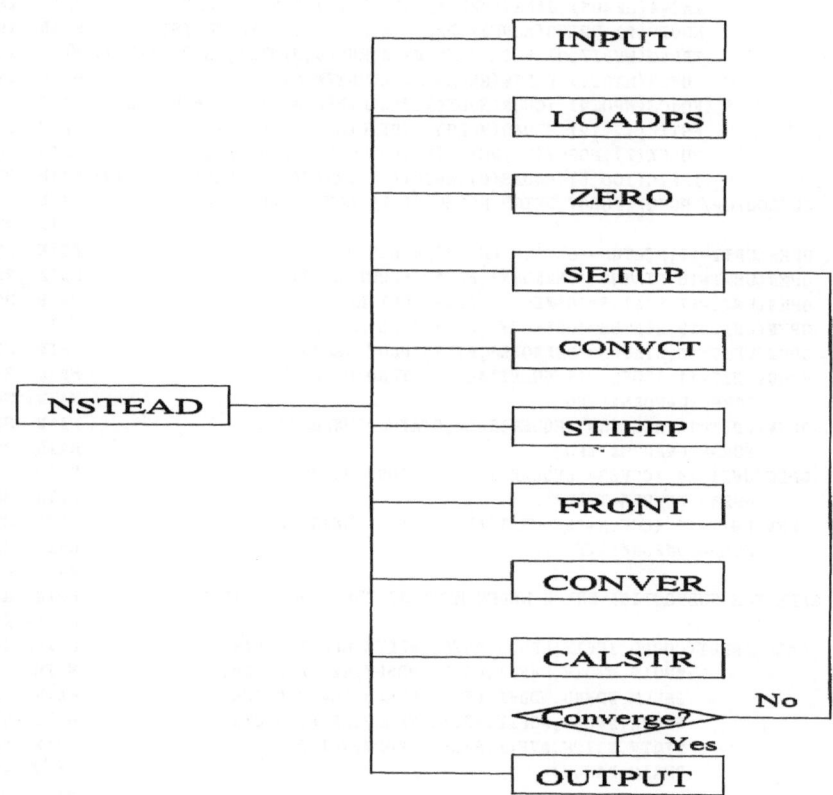

Figure A.1: NSTEAD Program structure

```
      PROGRAM NSTEAD                                                      MAIN   1
C***********************************************************************  MAIN   2
C     PROGRAM FOR THE STEADY-STATE NON-NEWTONIAN FLOW                     MAIN   3
C     OF PLANE AND AXISYMMETRIC PROBLEMS USING 3,6 NODED                  MAIN   4
C     TRIANGULAR ELEMENTS WITH MIXED FORMULATIIONS                        MAIN   5
C                               BY   H.C. HUANG   1992                    MAIN   6
C***********************************************************************  MAIN   7
      PARAMETER (MELEM=2000,MPOIN=4000,MFRON=1999,MBUFA=100,              MAIN   8
     .          MMATS=5,MTOTG=7*MELEM,NDOFN=2,MVFIX=999,NPROP=8,          MAIN   9
     .          MEVAB=NDOFN*6,MTOTV=MPOIN*NDOFN,                          MAIN  10
     .          MSTIF=MFRON*(MFRON+1)/2)                                  MAIN  11
C                                                                         MAIN  12
      IMPLICIT REAL*8 ( A-H, O-Z )                                        MAIN  13
C                                                                         MAIN  14
      DIMENSION ASDIS(MTOTV),COORD(MPOIN,2),ELOAD(MELEM,12),              MAIN  15
     .          EQRHS(MBUFA),EQUAT(MFRON,MBUFA),FIXED(MTOTV),             MAIN  16
```

```
          .        IFFIX(MTOTV),LNODS(MELEM,9),LOCEL(12),MATNO(MELEM),     MAIN  17
          .        NACVA(MFRON),NAMEV(MBUFA),NDEST(12),NDFRO(MELEM),       MAIN  18
          .        NOUTP(2),NPIVO(MBUFA),PRESC(MVFIX,3),PROPS(MMATS,8),    MAIN  19
          .        TREAC(MVFIX,2),VECRV(MFRON),STRSG(4,MTOTG),VELOC(MTOTV),MAIN  20
          .        TOFOR(MTOTV),PRESE(MELEM,3),EFFST(MTOTG),               MAIN  21
          .        FORCE(MPOIN),ICOUN(MPOIN),MARKP(MELEM,9),EPSTN(MTOTG),  MAIN  22
          .        ESTIF(12,12),GLOAD(MFRON),GSTIF(MSTIF),NOFIX(MVFIX),    MAIN  23
          .        POSGX(7),POSGY(7),WEIGP(7),POSEG(3),WEIEG(3),           MAIN  24
          .        IEFIC(100,5),BGUAS(6),BWEGT(6),TREAU(2),RLOAD(MELEM,12) MAIN  25
      COMMON/GP/ POSGX,POSGY,WEIGP,POSEG,WEIEG,BGUAS,BWEGT                 MAIN  26
C                                                                         MAIN  27
      OPEN(UNIT=15,STATUS='OLD',FILE='EPIN.DAT')                          MAIN  28
      OPEN(UNIT=16,STATUS='UNKNOWN',FILE='EPOUT.RES')                     MAIN  29
      OPEN(UNIT=17,STATUS='UNKNOWN',FILE='OUT.RES')                       MAIN  30
      OPEN(UNIT=19,STATUS='UNKNOWN',FILE='VEL.RES')                       MAIN  31
      OPEN(UNIT=36,STATUS='UNKNOWN',FILE='PLOT.RES')                      MAIN  32
      OPEN(UNIT=11,ACCESS='SEQUENTIAL',STATUS='UNKNOWN',                  MAIN  33
          .     FORM='UNFORMATTED')                                       MAIN  34
      OPEN(UNIT=12,ACCESS='SEQUENTIAL',STATUS='UNKNOWN',                  MAIN  35
          .     FORM='UNFORMATTED')                                       MAIN  36
      OPEN(UNIT=14,ACCESS='SEQUENTIAL',STATUS='UNKNOWN',                  MAIN  37
          .     FORM='UNFORMATTED')                                       MAIN  38
      OPEN(UNIT=18,ACCESS='SEQUENTIAL',STATUS='UNKNOWN',                  MAIN  39
          .     FORM='UNFORMATTED')                                       MAIN  40
C                                                                         MAIN  41
C*** CALL THE SUBROUTINE WHICH READS MOST OF THE PROBLEM DATA             MAIN  42
C                                                                         MAIN  43
      CALL INPUT(COORD,IFFIX,LNODS,MATNO,MELEM,MFRON,MMATS,               MAIN  44
          .           MPOIN,MTOTV,MVFIX,POSGX,POSGY,NNODP,NCONV,          MAIN  45
          .           NBELM,NDFRO,NDOFN,NELEM,NEVAB,NGAUS,NLINR,          MAIN  46
          .           NPOWR,NNODE,NOFIX,NPOIN,NPROP,NSTR1,NTOTG,          MAIN  47
          .           NTOTV,NTYPE,NVFIX,PRESC,PROPS,WEIGP,IEFIC,          MAIN  48
          .           PFRIC,RAMUV)                                        MAIN  49
C                                                                         MAIN  50
C*** CALL THE SUBROUTINE WHICH COMPUTES THE CONSISTENT LOAD VECTORS       MAIN  51
C    FOR EACH ELEMENT AFTER READING THE RELEVANT INPUT DATA               MAIN  52
C                                                                         MAIN  53
      CALL LOADPS(COORD,LNODS,MATNO,MELEM,MMATS,MPOIN,NELEM,              MAIN  54
          .            NEVAB,NGAUS,NNODE,NPOIN,NTYPE,POSGX,POSGY,         MAIN  55
          .            PROPS,RLOAD,WEIGP,NDOFN,                           MAIN  56
          .            POSEG,WEIEG)                                       MAIN  57
C                                                                         MAIN  58
C*** INITIALISE CERTAIN ARRAYS                                            MAIN  59
C                                                                         MAIN  60
      CALL ZERO(MELEM,MEVAB,MTOTG,MTOTV,NDOFN,NELEM,                      MAIN  61
          .           NEVAB,NSTR1,NTOTG,EFFST,EPSTN,MPOIN,ICOUN,          MAIN  62
          .           NTOTV,NVFIX,STRSG,VELOC,FORCE,                      MAIN  63
          .           TREAC,MVFIX,MMATS,PROPS,MARKP)                      MAIN  64
C                                                                         MAIN  65
C*** READ DATA FOR ITERATIONS                                             MAIN  66
C                                                                         MAIN  67
      CALL INDATA(MITER,NOUTP,TOLER)                                      MAIN  68
C                                                                         MAIN  69
C*** LOOP OVER EACH ITERATION                                             MAIN  70
C                                                                         MAIN  71
      DO 50 IITER = 1,MITER                                               MAIN  72
      PRINT *, 'IITER=',IITER                                             MAIN  73
```

```
C                                                               MAIN  74
C*** CALL ROUTINE WHICH SET UP BOUDARY CONDITIONS               MAIN  75
C                                                               MAIN  76
      CALL SETUP(FIXED,IITER,MVFIX,NDOFN,NVFIX,                 MAIN  77
     .           MTOTV,NTOTV,PRESC,NOFIX)                       MAIN  78
C                                                               MAIN  79
      CALL CONVCF(COORD,LNODS,MATNO,MELEM,MMATS,MPOIN,NELEM,    MAIN  80
     .            NEVAB,NGAUS,NNODE,NPOIN,NTYPE,POSGX,POSGY,    MAIN  81
     .       -    PROPS,ELOAD,WEIGP,NDOFN,VELOC,MTOTV,RLOAD,    MAIN  82
     .            IITER,NCONV)                                  MAIN  83
C                                                               MAIN  84
C*** CHECK WHETHER A NEW EVALUATION OF THE STIFFNESS MATRIX IS REQUIRED MAIN  85
C                                                               MAIN  86
      CALL STIFFP(COORD,EFFST,LNODS,MATNO,NPOWR,                MAIN  87
     .            MEVAB,MMATS,MPOIN,NELEM,NEVAB,NGAUS,NNODE,ESTIF, MAIN  88
     .            NSTR1,POSGX,POSGY,PROPS,WEIGP,MELEM,MTOTG,NLINR, MAIN  89
     .            NTYPE,NBELM,IITER,EPSTN,PRESC,MVFIX,NOFIX,NVFIX, MAIN  90
     .            MTOTV,VELOC,IEFIC,BGUAS,BWEGT,PFRIC,RAMUV,NNODP) MAIN  91
C                                                               MAIN  92
C*** SOLVE EQUATIONS                                            MAIN  93
C                                                               MAIN  94
      CALL FRONT(ASDIS,ELOAD,EQRHS,EQUAT,ESTIF,FIXED,IFFIX,IITER, MAIN  95
     .           GLOAD,GSTIF,LOCEL,LNODS,MBUFA,MELEM,MEVAB,MFRON, MAIN  96
     .           MSTIF,MTOTV,MVFIX,NACVA,NAMEV,NDEST,NDOFN,NELEM, MAIN  97
     .           NEVAB,NNODE,NPIVO,NPOIN,NTOTV,VELOC,TREAC,VECRV) MAIN  98
C                                                               MAIN  99
C*** CHECK FOR CONVERGENCE                                      MAIN 100
C                                                               MAIN 101
      CALL CONVER(ASDIS,VELOC,MTOTV,NTOTV,IITER,NCHEK,PVALU,    MAIN 102
     .            TOLER,MVFIX,NVFIX,NOFIX,PRESC)                MAIN 103
C                                                               MAIN 104
C*** CALCULATE  STRAIN RATES AND STRESSES                       MAIN 105
C                                                               MAIN 106
      CALL CALSTR(VELOC,COORD,EFFST,LNODS,NPOWR,                MAIN 107
     .            MATNO,MELEM,MMATS,MPOIN,MTOTG,MTOTV,NDOFN,    MAIN 108
     .            NELEM,NEVAB,NGAUS,NNODE,NSTR1,NTYPE,PROPS,    MAIN 109
     .            NBELM,STRSG,WEIGP,EPSTN,NPOIN,IITER,NLINR,    MAIN 110
     .            POSGX,POSGY,PRESE,NNODP,MITER,NCHEK)          MAIN 111
      TREAU(1)=0.0                                              MAIN 112
      TREAU(2)=0.0                                              MAIN 113
      DO 45 IVFIX=1,NVFIX                                       MAIN 114
      DO 45 IDOFN=1,2                                           MAIN 115
      IF(PRESC(IVFIX,IDOFN).NE.0.0)                            MAIN 116
     .TREAU(IDOFN)=TREAU(IDOFN)+TREAC(IVFIX,IDOFN)             MAIN 117
   45 CONTINUE                                                 MAIN 118
      WRITE(16,*)                                               MAIN 119
      WRITE(16,*) 'TOTAL FORCES ON THE RAM WITH PRESCRIBED VELOCITY' MAIN 120
      WRITE(16,*)                                               MAIN 121
      IF(TREAU(1).NE.0.0) WRITE(16,*) ' U-FORCE = ', TREAU(1)  MAIN 122
      IF(TREAU(2).NE.0.0) WRITE(16,*) ' V-FORCE = ', TREAU(2)  MAIN 123
C                                                               MAIN 124
      IF(TREAU(1).NE.0.0) PRINT *, ' U-FORCE = ', TREAU(1)     MAIN 125
      IF(TREAU(2).NE.0.0) PRINT *, ' V-FORCE = ', TREAU(2)     MAIN 126
      print *,'nchek= ',nchek                                  MAIN 127
      IF(NCHEK.EQ.0.OR.NCHEK.EQ.999) GO TO 101                 MAIN 128
   50 CONTINUE                                                 MAIN 129
  101 CONTINUE                                                 MAIN 130
```

```
C                                                                MAIN 131
C*** OUTPUT RESULTS IF REQUIRED                                  MAIN 132
C                                                                MAIN 133
      CALL OUTPUT(IITER,MTOTG,MTOTV,MVFIX,NELEM,NGAUS,NOFIX,NOUTP, MAIN 134
     .           NPOIN,NVFIX,STRSG,VELOC,TREAC,EPSTN,NTYPE,NCHEK,   MAIN 135
     .           MELEM,MARKP,PRESE,PRESC,NNODE,LNODS,FORCE,MPOIN,   MAIN 136
     .           MATNO,ICOUN,NTOTV,NNODP)                           MAIN 137
C                                                                MAIN 138
      STOP                                                       MAIN 139
      END                                                        MAIN 140
```

A.4   Input Instructions

The input subroutine is listed here followed by complete instructions on preparing the input data file.

```
      SUBROUTINE INPUT(COORD,IFFIX,LNODS,MATNO,MELEM,MFRON,MMATS, INPU  1
     .                MPOIN,MTOTV,MVFIX,POSGX,POSGY,NNODP,NCONV,   INPU  2
     .                NBELM,NDFRO,NDOFN,NELEM,                     INPU  3
     .                NEVAB,NGAUS,NLINR,NPOWR,                     INPU  4
     .                NNODE,NOFIX,NPOIN,NPROP,NSTR1,               INPU  5
     .                NTOTG,NTOTV,NTYPE,NVFIX,PRESC,PROPS,         INPU  6
     .                WEIGP,IEFIC,PFRIC,RAMUV)                     INPU  7
C************************************************************************ INPU  8
C                                                                INPU  9
C**** THIS SUBROUTINE ACCEPTS MOST OF THE INPUT DATA             INPU 10
C                                                                INPU 11
C************************************************************************ INPU 12
C     INSERT DOUBLE                                              INPU 13
C                                                                INPU 14
      IMPLICIT REAL*8 ( A-H, O-Z )                               INPU 15
C                                                                INPU 16
      DIMENSION COORD(MPOIN,2),IFFIX(MTOTV),LNODS(MELEM,9),      INPU 17
     .          MATNO(MELEM),NDFRO(MELEM),POSGX(7),POSGY(7),     INPU 18
     .          NOFIX(MVFIX),PRESC(MVFIX,3),IEFIC(100,5),        INPU 19
     .          PROPS(MMATS,NPROP),TITLE(12),WEIGP(7)            INPU 20
      REWIND 11                                                  INPU 21
      REWIND 12                                                  INPU 22
C     REWIND 3                                                   INPU 23
      REWIND 14                                                  INPU 24
      REWIND 18                                                  INPU 25
      READ(15,920)  TITLE                                       INPU 26
      WRITE(16,920) TITLE                                       INPU 27
  920 FORMAT(12A6)                                              INPU 28
C                                                                INPU 29
C*** READ THE FIRST DATA CARD, AND ECHO IT IMMEDIATELY.         INPU 30
C                                                                INPU 31
      READ(15,*) NPOIN,NELEM,NVFIX,NTYPE,NNODE,NMATS,NGAUS,     INPU 32
     .NNODP,NBELM,NCONV,NSTRE,NLINR,NPOWR                       INPU 33
  900 FORMAT(13I5)                                              INPU 34
      NEVAB=NDOFN*NNODE                                         INPU 35
      NSTR1=NSTRE+1                                             INPU 36
```

```
      IF(NTYPE.EQ.3) NSTR1=NSTRE                                     INPU 37
      NTOTV=NPOIN*NDOFN                                              INPU 38
      NGAU2=NGAUS                                                    INPU 39
      NTOTG=NELEM*NGAU2                                              INPU 40
      WRITE(16,901)NPOIN,NELEM,NVFIX,NTYPE,NNODE,NMATS,NGAUS,NEVAB,  INPU 41
     .NNODP,NBELM,NCONV,NSTRE,NLINR,NPOWR                            INPU 42
  901 FORMAT(//8H NPOIN =,I4,4X,8H NELEM =,I4,4X,8H NVFIX =,I4,4X,   INPU 43
     . 8H NTYPE =,I4,4X,8H NNODE =,I4,//                             INPU 44
     . 8H NMATS =,I4,4X,8H NGAUS =,I4,                               INPU 45
     .             4X,8H NEVAB =,I4,4X,8H NNODP =,I4//               INPU 46
     . 8H NBELM =,I4,4X,8H NCONV =,I4,4X,8H NSTRE =,I4,4X,           INPU 47
     . 8H NLINR =,I4,4X,8H NPOWR =,I4)                               INPU 48
      CALL      CHECK1(NDOFN,NELEM,NGAUS,NMATS,NNODE,NPOIN,          INPU 49
     .                 NSTRE,NTYPE,NVFIX,NBELM,NNODP)                INPU 50
C                                                                    INPU 51
C*** READ THE ELEMENT NODAL CONNECTIONS, AND THE PROPERTY NUMBERS.   INPU 52
C                                                                    INPU 53
      WRITE(16,902)                                                  INPU 54
  902 FORMAT(//8H ELEMENT,3X,8HPROPERTY,6X,12HNODE NUMBERS)          INPU 55
      DO 2 IELEM=1,NELEM                                             INPU 56
      READ(15,*)NUMEL,MATNO(NUMEL),(LNODS(NUMEL,INODE),INODE=1,NNODE) INPU 57
    2 WRITE(16,903)NUMEL,MATNO(NUMEL),(LNODS(NUMEL,INODE),INODE=1,NNODE)INPU 58
  903 FORMAT(1X,I5,I9,6X,8I5)                                        INPU 59
C                                                                    INPU 60
C*** ZERO ALL THE NODAL COORDINATES, PRIOR TO READING SOME OF THEM.  INPU 61
C                                                                    INPU 62
      DO 4 IPOIN=1,NPOIN                                             INPU 63
      DO 4 IDIME=1,2                                                 INPU 64
    4 COORD(IPOIN,IDIME)=0.0                                         INPU 65
C                                                                    INPU 66
C*** READ SOME NODAL COORDINATES, FINISHING WITH THE LAST NODE OF ALL. INPU 67
C                                                                    INPU 68
      WRITE(16,904)                                                  INPU 69
  904 FORMAT(//5H NODE,10X,1HX,10X,1HY)                              INPU 70
    6 READ(15,*) IPOIN,(COORD(IPOIN,IDIME),IDIME=1,2)                INPU 71
  905 FORMAT(I5,6F10.5)                                              INPU 72
      IF(IPOIN.NE.NPOIN) GO TO 6                                     INPU 73
C                                                                    INPU 74
C*** WRITE OUT COORDINATES                                           INPU 75
C                                                                    INPU 76
      DO 10 IPOIN=1,NPOIN                                            INPU 77
   10 WRITE(16,906) IPOIN,(COORD(IPOIN,IDIME),IDIME=1,2)             INPU 78
  906 FORMAT(1X,I5,3F10.3)                                           INPU 79
C                                                                    INPU 80
C*** READ THE FIXED VALUES.                                          INPU 81
C                                                                    INPU 82
      WRITE(16,907)                                                  INPU 83
  907 FORMAT(//5H NODE,6X,4HCODE,6X,12HFIXED VALUES)                 INPU 84
      DO 7 ITOTV=1,NTOTV                                             INPU 85
    7 IFFIX(ITOTV)=0                                                 INPU 86
      DO 8 IVFIX=1,NVFIX                                             INPU 87
      READ(15,*)NOFIX(IVFIX),IFPRE,(PRESC(IVFIX,IDOFN),IDOFN=1,3)    INPU 88
      WRITE(16,908)NOFIX(IVFIX),IFPRE,(PRESC(IVFIX,IDOFN),IDOFN=1,3) INPU 89
      NLOCA=(NOFIX(IVFIX)-1)*NDOFN                                   INPU 90
      IFDOF=10**(NDOFN-1)                                            INPU 91
      DO 8 IDOFN=1,NDOFN                                             INPU 92
      NGASH=NLOCA+IDOFN                                              INPU 93
```

```
      IF(IFPRE.LT.IFDOF) GO TO 8                                    INPU  94
      IFFIX(NGASH)=1                                                INPU  95
      IFPRE=IFPRE-IFDOF                                             INPU  96
    8 IFDOF=IFDOF/10                                                INPU  97
  908 FORMAT(1X,I4,5X,I5,5X,5F10.6)                                 INPU  98
C                                                                   INPU  99
C*** READ THE AVAILABLE SELECTION OF ELEMENT PROPERTIES.           INPU 100
C                                                                   INPU 101
      WRITE(16,910)                                                 INPU 102
  910 FORMAT(//7H NUMBER,6X,18HELEMENT PROPERTIES)                  INPU 103
      DO 18 IMATS=1,NMATS                                           INPU 104
      READ(15,*) NUMAT                                              INPU 105
      READ(15,*) (PROPS(NUMAT,IPROP),IPROP=1,NPROP)                 INPU 106
  930 FORMAT(8F10.5)                                                INPU 107
   18 WRITE(16,911) NUMAT,(PROPS(NUMAT,IPROP),IPROP=1,NPROP)        INPU 108
  911 FORMAT(1X,I4,3X,8E14.6)                                       INPU 109
      WRITE(16,*)                                                   INPU 110
      READ(15,*) PFRIC,RAMUV                                        INPU 111
      WRITE(16,*) 'FRICTION COEEF. = ',PFRIC                        INPU 112
      WRITE(16,*) 'RAM SPEED = ',RAMUV                              INPU 113
      WRITE(16,*)                                                   INPU 114
      IF(NBELM.NE.0) THEN                                           INPU 115
      WRITE(16,*) 'INFORMATION FOR FRICTIONAL ELEMENTS'             INPU 116
      WRITE(16,*)                                                   INPU 117
      DO 25 IBELM=1,NBELM                                           INPU 118
      READ(15,*) (IEFIC(IBELM,IDUMY),IDUMY=1,5)                     INPU 119
   25 WRITE(16,900)(IEFIC(IBELM,IDUMY),IDUMY=1,5)                   INPU 120
      END IF                                                        INPU 121
C                                                                   INPU 122
C*** SET UP GAUSSIAN INTEGRATION CONSTANTS                          INPU 123
C                                                                   INPU 124
      CALL      GAUSSQ(NGAUS,POSGX,POSGY,WEIGP)                     INPU 125
      CALL      CHECK2(COORD,IFFIX,LNODS,MATNO,MELEM,MFRON,MPOIN,MTOTV,INPU 126
     .              MVFIX,NDFRO,NDOFN,NELEM,NMATS,NNODE,NOFIX,NPOIN,INPU 127
     .              NVFIX)                                          INPU 128
C                                                                   INPU 129
C*** OUTPUT INFORMATATION FOR PLOTING                               INPU 130
C                                                                   INPU 131
      WRITE(36,*) 'NO.-OF-BACKGROUND-POINTS-AND-ELEMENTS'           INPU 132
      WRITE(36,352) NPOIN, NELEM, NNODE, -1                         INPU 133
      WRITE(36,*) 'COORDINATES'                                     INPU 134
      DO 7001 INP=1,NPOIN                                           INPU 135
 7001 WRITE(36,354) INP,COORD(INP,1),COORD(INP,2),0.999,1.0,0.0,0.0 INPU 136
      WRITE(36,*) 'ELEMENT-CONNECTIVITY'                            INPU 137
      DO 7002 IE=1,NELEM                                            INPU 138
 7002 WRITE(36,86) IE,MATNO(IE),(LNODS(IE,I1),I1=1,NNODE,1)         INPU 139
C                                                                   INPU 140
   86 FORMAT(11I5)                                                  INPU 141
  352 FORMAT(4I5)                                                   INPU 142
  354 FORMAT(I5,6E12.4)                                             INPU 143
      RETURN                                                        INPU 145
      END                                                           INPU 146
```

All the input data required by the program is read in subroutine **INPUT** listed above from the data file **EPIN.DAT**. The details of the input procedure are described below. The variables and arrays to be input are written in bold face. For convenience the variables are described following each read command, although they were defined in previous sections.

1 Control data

    1.1 READ(NIN,'(80A1)')**TITLE**

    1.2 READ(NIN,∗)**NPOIN, NELEM, NVFIX, NTYPE, NNODE, NMATS, NGAUS, NNODP, NBELM, NCONV, NSTRE, NLINR, NPOWR**

  where

  NPOIN: Number of nodes,

  NELEM: Number of elements,

  NVFIX: Number of boundary nodes,

  NTYPE: 1 = plane stress only if nlinr=0; 2 = plane problem; 3 = axisymmetric problem,

  NNODE: Number of nodes per element,

  NMATS: Number of material sets,

  NGAUS: Number of gauss stations per element,

  NNODP: Number of pressure nodes per element,

  NBELM: Number of elements on slip boundaries,

  NCONV: 1 - adding convection term; 0 - no convection term,

  NSTRE: Number of stresses, plane problem using 3; axisymmetric case using 4,

  NLINR: 0 for linear elastic and 1 for non-linear problems,

  NPOWR: 1 = for power law flow; 0 for plastic flow.

2 Mesh data

    2.1 READ(NIN,∗) **NUMEL, MATNO** (NUMEL), (**LNODS** (NUMEL,INODE), INODE=1, NNODE)

    (loop over NELEM elements)

  where

  NUMEL: Current element number,

MATNO (NUMEL): material set for current element NUMEL,

LNODS (NUMEL,INODE),INODE=1,NNODE): element nodal connections,

2.2 READ(NIN,∗) **IPOIN**, (**COORD** (IPOIN, IDIME), IDIME = 1,2)

(loop over NPOIN nodes)

where

IPOIN: Current nodal number,

COORD (IPOIN,IDIME),IDIME=1,2): coordinates (IDIME = 1 for x and 2 for y axies in plane problem; 1 for r and 2 for z axies in axialsymmetric case.)

3 Boundary conditions

3.1 READ(NIN,∗) **NOFIX** (IVFIX), **IFPRE**, (**PRESC** (IVFIX, IDOFN), IDOFN=1, 3)

(loop over NVFIX boundary nodes)

where

NOFIX (IVFIX): Boundary nodal numbers,

IFPRE: indicator for boundary conditions (IFPRE = 1 for x or r direction with fixed value; 10 for y or z direction with fixed value; 11 - x and and directions or r and z directions both fixed with given values)

PRESC (IVFIX,I): Fixed boundary velocities (If PRESC (IVFIX,3) = 0 then PRESC (IVFIX,1) and PRESC (IVFIX,2) refer to fixed velocities in 1 and 2 directions, otherwise PRESC (IVFIX,1) and PRESC (IVFIX,2) refer to the fixed values in tangent and normal directions with PRESC (IVFIX,3), the given angle between boundary edge and 1-axis.)

4 Material properties

4.1 READ(NIN,∗) **NUMAT**

4.2 READ(NIN,∗) (**PROPS** (NUMAT,IPROP), IPROP = 1, NPROP)

(loop over NMATS material sets)

where

NUMAT: Current material set

NPROP: Number of material properties (given as 8)

PROPS (NUMAT,1): E - Young's modulus

PROPS (NUMAT,2): ν - Poisson's ratio

PROPS (NUMAT,3): thickness

PROPS (NUMAT,4): p or n of the power law index

PROPS (NUMAT,5): μ_o constant in power law flow

PROPS (NUMAT,6): Hardening parameter (H') for linear strain hardening

PROPS (NUMAT,7): Pressure penalty (1e+08)

PROPS (NUMAT,8): Mass density ρ.

5 Friction boundary data

  5.1 READ(NIN,*) **PFRIC,RAMUV**

  where

  PFRIC: Friction coefficient

  RAMUV: Ram speed

Option data for slip boundaries (if NBELM (nuber of boudary slip elements) > 0)

  5.2* READ(NIN,*) (**IEFIC** (IBELM,IDUMY),IDUMY=1,5)

    (loop over nbelm times)

  where

  IEFIC (IBELM,1): Current element number

  IEFIC (IBELM,2): 1st node on the friction edge

  IEFIC (IBELM,3): 2nd node on the friction edge

  IEFIC (IBELM,4): 3rd node on the friction edge

  IEFIC (IBELM,5): ith freedom for the friction edge

6 Title card

  6.1 READ(NIN,'(80a1)') **TITLE**

7 Data card for loads

    7.1 READ(NIN,∗) **IPLOD,IGRAV,IEDGE**

    where

    IPLOD: Number of point loads

    ·IGRAV: Indicator; > 0 for gravity

    IEDGE: Number of edge with tractions

Option cards for point loads (if IPLOAD (point loads) > 0)

    7.2∗ READ(NIN,∗) **LODPT**, (**POINT** (IDOFN), IDOFN = 1,2)

    (loop over IPLOAD times)

    where

    LODPT: Node number where point load is applied

    POINT (IDOFN): Point loads in 1 and 2 directions

Option cards for gravity case (if IGRAV (gravity) > 0)

    7.3∗ read(NIN,∗) **THETA,GRAVY**

    where

    THETA: Gravity angle

    GRAVY: Gravitational constant

Option cards for tractions (if IEDGE > 0)

    7.4∗ READ(NIN,∗) **NEDGE**

    7.4.1∗ READ(NIN,∗) **NEASS**, (**NOPRS** (IODEG), IODEG = 1, NODEG)

    7.4.2∗ READ(NIN,∗) ((**PRESS** (IODEG, IDOFN), IDOFN = 1, 2), IODEG = 1, NODEG)

    (loop over NEDGE times for 7.4.1∗ and 7.4.2∗)

    NEDGE: Number of edges with traction loads

    NODEG: Number of nodes per side of an element

    NEASS: Current number of element with traction loads

    NOPRS: Store nodes at the edge with tractions

    PRESS: Normal and tangetial distributed loads at IODEGth node

8 Parameter card

8.1 READ(NIN,*) **FACTO, TOLER, MITER, NOUTP(1), NOUTP(2)**

where

FACTO: Dumy factor

TOLER: Convergence tolerance

MITER: Max. no. of iterations

NOUTP (1): Initial output parameter (ref. below)

NOUTP (2): Final output parameter (ref. below)

0 = No output

1 = Output velocities

2 = Output velocities + reactions

3 = Output velocities + reactions + stresses and strains

A.5 Element Stiffness Calculations

Element stiffness matrices are calculated in the following routine.

```
      SUBROUTINE STIFFP(COORD,EFFST,LNODS,MATNO,NPOWR,MEVAB,MMATS,     STIF   1
     .           MPOIN,NELEM,NEVAB,NGAUS,NNODE,ESTIF,                   STIF   2
     .           NSTR1,POSGX,POSGY,PROPS,WEIGP,MELEM,MTOTG,             STIF   3
     .           NLINR,NTYPE,NBELM,IITER,EPSTN,PRESC,MVFIX,             STIF   4
     .           NOFIX,NVFIX,MTOTV,VELOC,IEFIC,BGUAS,BWEGT,             STIF   5
     .           PFRIC,RAMUV,NNODP)                                     STIF   6
C*******************************************************************     STIF   7
C                                                                       STIF   8
C**** THIS SUBROUTINE EVALUATES THE STIFFNESS MATRIX FOR EACH ELEMENT    STIF   9
C     IN TURN                                                           STIF  10
C                                                                       STIF  11
C*******************************************************************     STIF  12
C     INSERT DOUBLE                                                     STIF  13
C                                                                       STIF  14
      IMPLICIT REAL*8 ( A-H, O-Z )                                      STIF  15
C                                                                       STIF  16
      DIMENSION BMATX(4,12),CARTD(2,9),COORD(MPOIN,2),                  STIF  17
     .          DERIV(2,9),DEVIA(4),ELVEL(2,9),                        STIF  18
     .          ELCOD(2,9),EFFST(MTOTG),LNODS(MELEM,9),                STIF  19
     .          MATNO(MELEM),POSGX(7),POSGY(7),PROPS(MMATS,8),SHAPE(9), STIF  20
     .          WEIGP(7),VELOC(MTOTV),CMATX(3,3),                      STIF  21
     .          GPCOD(2,9),ESTIF(MEVAB,MEVAB),                         STIF  22
     .          EPSTN(MTOTG),PRESC(MVFIX,3),NOFIX(MVFIX),QMATX(12,3),  STIF  23
     .          IEFIC(100,5),BGUAS(6),BWEGT(6)                         STIF  24
```

```
      TWOPI=6.283185308                                   STIF  25
      REWIND 11                                           STIF  26
      KGAUS=0                                             STIF  27
C                                                         STIF  28
C*** LOOP OVER EACH ELEMENT                               STIF  29
C                                                         STIF  30
      DO 70 IELEM=1,NELEM                                 STIF  31
      LPROP=MATNO(IELEM)                                  STIF  32
C                                                         STIF  33
C*** EVALUATE THE COORDINATES OF THE ELEMENT NODAL POINTS STIF  34
C                                                         STIF  35
      DO 10 INODE=1,NNODE                                 STIF  36
      LNODE=IABS(LNODS(IELEM,INODE))                      STIF  37
      NPOSN=(LNODE-1)*2                                   STIF  38
      DO 10 IDIME=1,2                                     STIF  39
      NPOSN=NPOSN+1                                       STIF  40
      ELCOD(IDIME,INODE)=COORD(LNODE,IDIME)               STIF  41
   10 ELVEL(IDIME,INODE)=VELOC(NPOSN)                     STIF  42
      YOUNG=PROPS(LPROP,1)                                STIF  43
      POISS=PROPS(LPROP,2)                                STIF  44
      THICK=PROPS(LPROP,3)                                STIF  45
      PINDX=PROPS(LPROP,4)                                STIF  46
      UNIAX=PROPS(LPROP,5)                                STIF  47
      HARDS=PROPS(LPROP,6)                                STIF  48
      PENAT=PROPS(LPROP,7)                                STIF  49
      IF(NLINR.EQ.0) THEN                                 STIF  50
      VISCO=0.5*YOUNG/(1.0+POISS)                         STIF  51
      POIS2=2.0*POISS                                     STIF  52
      IF(NTYPE.EQ.1) POIS2=POISS                          STIF  53
      DLUMD=2.0*VISCO*POISS/(1.0-POIS2)                   STIF  54
      END IF                                              STIF  55
C                                                         STIF  56
C*** INITIALIZE THE ELEMENT STIFFNESS MATRICES            STIF  57
C                                                         STIF  58
      DO 15 INODP=1,NNODP                                 STIF  59
      DO 15 JNODP=1,NNODP                                 STIF  60
   15 CMATX(INODP,JNODP)=0.0                              STIF  61
C                                                         STIF  62
      DO 20 IEVAB=1,NEVAB                                 STIF  63
      DO 25 INODP=1,NNODP                                 STIF  64
   25 QMATX(IEVAB,INODP)=0.0                              STIF  65
      DO 20 JEVAB=1,NEVAB                                 STIF  66
   20 ESTIF(IEVAB,JEVAB)=0.0                              STIF  67
      KGASP=0                                             STIF  68
C                                                         STIF  69
C*** ENTER LOOPS FOR AREA NUMERICAL INTEGRATION           STIF  70
C                                                         STIF  71
      DO 50 IGAUS=1,NGAUS                                 STIF  72
      EXISP=POSGX(IGAUS)                                  STIF  73
      ETASP=POSGY(IGAUS)                                  STIF  74
      KGASP=KGASP+1                                       STIF  75
      KGAUS=KGAUS+1                                       STIF  76
C                                                         STIF  77
C*** EVALUATE THE SHAPE FUNCTIONS,ELEMENTAL VOLUME,ETC.   STIF  78
C                                                         STIF  79
      CALL      SFR2(DERIV,ETASP,EXISP,NNODE,SHAPE)       STIF  80
      CALL      JACOB2(CARTD,DERIV,DJACB,ELCOD,GPCOD,IELEM,KGASP,  STIF  81
```

```
                            NNODE,SHAPE)                               STIF  82
          DVOLU=DJACB*WEIGP(IGAUS)                                     STIF  83
          IF(NTYPE.EQ.3) DVOLU=DVOLU*TWOPI*GPCOD(1,KGASP)             STIF  84
          IF(THICK.NE.0.0) DVOLU=DVOLU*THICK                          STIF  85
C                                                                     STIF  86
C*** EVALUATE THE B MATRICES                                          STIF  87
C                                                                     STIF  88
          CALL BMATPS(BMATX,CARTD,NNODE,SHAPE,GPCOD,NTYPE,KGASP)      STIF  89
C                                                                     STIF  90
          IF(NLINR.NE.0) THEN                                        STIF  91
          EFFGP=EPSTN(KGAUS)                                          STIF  92
          YIELD=UNIAX+HARDS*EFFGP                                     STIF  93
          IF(EFFGP.LE.0.0) EFFGP=0.1E-05                             STIF  94
          IF(NPOWR.EQ.0) VISCO=YIELD/EFFGP/3.0                       STIF  95
          IF(NPOWR.EQ.1) VISCO=YIELD*EFFGP**(PINDX-1.0)              STIF  96
          DLUMD=PENAT*VISCO                                           STIF  97
          END IF                                                      STIF  98
C                                                                     STIF  99
          DCONT=2.0*VISCO*DVOLU                                       STIF 100
C                                                                     STIF 101
C*** CALCULATE THE ELEMENT STIFFNESSES                                STIF 102
C                                                                     STIF 103
          DO 30 IEVAB=1,NEVAB                                         STIF 104
          DO 30 JEVAB=IEVAB,NEVAB                                     STIF 105
          DO 30 ISTRE=1,NSTR1                                         STIF 106
       30 ESTIF(IEVAB,JEVAB)=ESTIF(IEVAB,JEVAB)+BMATX(ISTRE,IEVAB)*   STIF 107
         .BMATX(ISTRE,JEVAB)*DCONT                                    STIF 108
C                                                                     STIF 109
C*** TO FORM Q-MATX                                                   STIF 110
C                                                                     STIF 111
          SHAPE(1)=1.0                                                STIF 112
          IF(NNODP.EQ.3) CALL SFR2(DERIV,ETASP,EXISP,NNODP,SHAPE)     STIF 113
          DO 40 IEVAB=1,NEVAB                                         STIF 114
          DO 40 INODP=1,NNODP                                         STIF 115
          DO 40 ISTRE=1,3                                             STIF 116
       40 QMATX(IEVAB,INODP)=QMATX(IEVAB,INODP)+DVOLU*BMATX(ISTRE,IEVAB)* STIF 117
         .SHAPE(INODP)                                                STIF 118
C                                                                     STIF 119
C*** TO FORM C-MATX                                                   STIF 120
C                                                                     STIF 121
          DO 42 INODP=1,NNODP                                         STIF 122
          DO 42 JNODP=1,NNODP                                         STIF 123
       42 CMATX(INODP,JNODP)=CMATX(INODP,JNODP)+(DVOLU/DLUMD)*        STIF 124
         .SHAPE(INODP)*SHAPE(JNODP)                                   STIF 125
       50 CONTINUE                                                    STIF 126
C                                                                     STIF 127
C*** TO GET THE INVERSED C-MATRIX                                     STIF 128
C                                                                     STIF 129
          CALL INVERS(CMATX,NNODP,FAILD)                              STIF 130
           IF(FAILD.LT.0.)THEN                                        STIF 131
           PRINT *,' C-MATRIX INVERSING FAILED !!'                    STIF 132
           PRINT *,' CMATX(1,1),FAILD,IE=',CMATX(1,1),FAILD,IELEM     STIF 133
           WRITE(16,*)' C-MATRIX INVERSING FAILED !!'                 STIF 134
           STOP                                                       STIF 135
           END IF                                                     STIF 136
C                                                                     STIF 137
C*** ELIMNATE P-VARIABLES AND RECONSTRUCT STIFFNESS MATRIX.           STIF 138
```

```
C                                                              STIF 139
      DO 45 IEVAB=1,NEVAB                                      STIF 140
      DO 45 JEVAB=IEVAB,NEVAB                                  STIF 141
      DUMMY=0.0                                                STIF 142
      DO 48 INODP=1,NNODP                                      STIF 143
      DO 48 JNODP=1,NNODP                                      STIF 144
   48 DUMMY=DUMMY+QMATX(IEVAB,INODP)*CMATX(INODP,JNODP)*       STIF 145
     .QMATX(JEVAB,JNODP)                                       STIF 146
   45 ESTIF(IEVAB,JEVAB)=ESTIF(IEVAB,JEVAB)+DUMMY              STIF 147
C                                                              STIF 148
C*** CONSTRUCT THE LOWER TRIANGLE OF THE STIFFNESS MATRIX      STIF 149
C                                                              STIF 150
      DO 60 IEVAB=1,NEVAB                                      STIF 151
      DO 60 JEVAB=1,NEVAB                                      STIF 152
   60 ESTIF(JEVAB,IEVAB)=ESTIF(IEVAB,JEVAB)                    STIF 153
C                                                              STIF 154
C*** CHECK IF THE ELEMENT NODAL POINTS ON INCLINED BOUNDARY    STIF 155
C                                                              STIF 156
      DO 65 INODE=1,NNODE                                      STIF 157
      LNODE=IABS(LNODS(IELEM,INODE))                           STIF 158
      DO 68 IVFIX=1,NVFIX                                      STIF 159
      IF(LNODE.NE.NOFIX(IVFIX)) GO TO 68                       STIF 160
      IF(PRESC(IVFIX,3).EQ.0.0) GO TO 65                       STIF 161
      GO TO 75                                                 STIF 162
   68 CONTINUE                                                 STIF 163
      GO TO 65                                                 STIF 164
   75 DUMMY=PRESC(IVFIX,3)                                     STIF 165
      ANGLE=DUMMY*3.141592653589793D0/180.0D0                  STIF 166
      COSAP=DCOS(ANGLE)                                        STIF 167
      SINAP=DSIN(ANGLE)                                        STIF 168
      CALL TRANSE(ESTIF,INODE,COSAP,SINAP,MEVAB,NEVAB)         STIF 169
   65 CONTINUE                                                 STIF 170
C                                                              STIF 171
C*** CHECK IF THE ELEMENT IS ON FRICTIONAL EDGE.               STIF 172
C                                                              STIF 173
      IF(PFRIC.EQ.0.)GOTO 301                                  STIF 174
      DO 300 IBELM=1,NBELM                                     STIF 175
      KBONE=IEFIC(IBELM,1)                                     STIF 176
      IF(KBONE.EQ.IELEM)                                       STIF 177
     .CALL BUMTX(IBELM,NGAUS,MTOTG,NTYPE,PFRIC,IEFIC,ESTIF,    STIF 178
     .           POSGX,EFFST,ELCOD,ELVEL,RAMUV,BGUAS,BWEGT)    STIF 179
  300 CONTINUE                                                 STIF 180
  301 CONTINUE                                                 STIF 181
C                                                              STIF 182
C*** STORE THE STIFFNESS MATRIX,STRESS MATRIX AND SAMPLING POINT  STIF 183
C    COORDINATES FOR EACH ELEMENT ON DISC FILE                STIF 184
C                                                              STIF 185
      WRITE(11) ESTIF                                          STIF 186
   70 CONTINUE                                                 STIF 187
      RETURN                                                   STIF 188
      END                                                      STIF 189
```

A.6   Documented Examples

Two examples are considered in this section to show the preparation of input data and the results obtained using the steady state code NSTEAD. The first one is the general Newtonian flow in a plane channel with slip boundary (friction boundary conditions) and the second one is the flow over a step as has been described in the examples of chapter 4.

The input and result files for this two examples are listed at following subsections. To keep the print out of theses files within the page boundary, some format has been altered and spaces have been shortened.

A.6.1   Example 1: Flow in a Plane channel

The example chosen is a 2-D plane channel flow which is illustrated in Figure A.2 for a general non-Newtonian flow with power law index equal to 0.6. The geometry and boundary conditions are given in Figure A.2.a. The fluid is allowed to slide at the upper boundary with friction coefficient of 0.65 and normalised maximum speed of unity. The finite elemet mesh is shown in A.2.b with eight elements arranged at the slip boundary, while the calculated velocity profiles across the channel are given in A.2.c.

Input Data File

Contents of the **EPIN.DAT** for this example are as follows:

```
Appendix A example 1: friction f=0.65,p=0.6,parabolic inlet
289   128    64    2    6    1    7    3    8    0    3    1    1
        1     1    1   84   11   83   10   82
        2     1    1   86    2   85   11   84
        3     1    2   88   12   87   11   85
        4     1    2   90    3   89   12   88
        5     1    3   92   13   91   12   89
        6     1    3   94    4   93   13   92
        7     1    4   96   14   95   13   93
        8     1    4   98    5   97   14   96
```

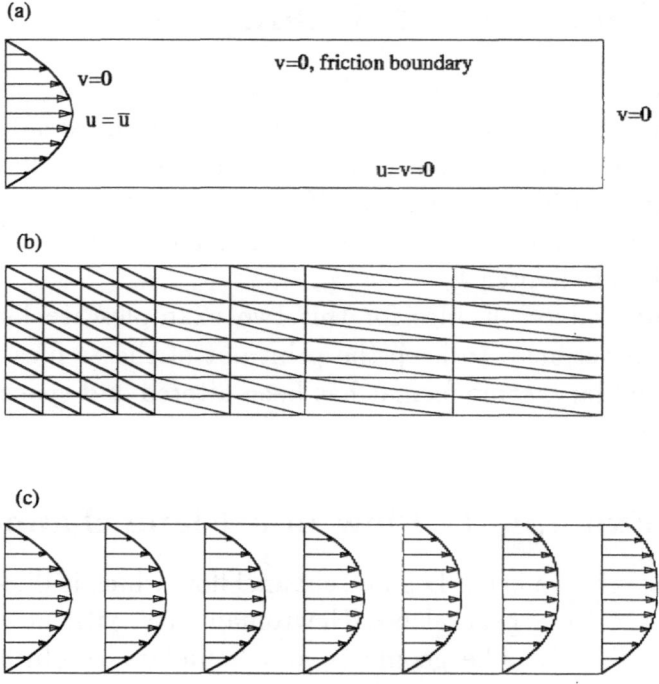

Figure A.2: Example 1: Flow in a plane channel, a) geomentry and boundary conditions; b) finite element mesh; c) velocity profiles.

. . .
. . .

115	1	65	270	75	269	74	268
116	1	65	244	66	271	75	270
117	1	66	273	76	272	75	271
118	1	66	247	67	274	76	273
119	1	67	276	77	275	76	274
120	1	67	250	68	277	77	276
121	1	68	279	78	278	77	277
122	1	68	253	69	280	78	279
123	1	69	282	79	281	78	280
124	1	69	256	70	283	79	282
125	1	70	285	80	284	79	283
126	1	70	259	71	286	80	285

127	1	71	288	81	287	80	286
128	1	71	262	72	289	81	288

1	40.00000	0.00000
2	40.00000	1.25000
3	40.00000	2.50000
4	40.00000	3.75000
5	40.00000	5.00000
6	40.00000	6.25000
7	40.00000	7.50000
8	40.00000	8.75000
9	40.00000	10.00000
10	30.00000	0.00000
11	30.00000	1.25000
12	30.00000	2.50000
13	30.00000	3.75000
14	30.00000	5.00000
15	30.00000	6.25000
16	30.00000	7.50000
17	30.00000	8.75000
18	30.00000	10.00000
19	20.00000	0.00000
20	20.00000	1.25000
21	20.00000	2.50000
22	20.00000	3.75000
23	20.00000	5.00000
24	20.00000	6.25000
25	20.00000	7.50000
26	20.00000	8.75000
27	20.00000	10.00000
28	15.00000	0.00000
29	15.00000	1.25000
30	15.00000	2.50000
31	15.00000	3.75000
32	15.00000	5.00000
33	15.00000	6.25000
34	15.00000	7.50000
35	15.00000	8.75000
36	15.00000	10.00000
37	10.00000	0.00000
38	10.00000	1.25000

```
 39   10.00000    2.50000


 ...
 ...


280    1.25000    6.25000
281    0.00000    6.87500
282    1.25000    6.87500
283    1.25000    7.50000
284    0.00000    8.12500
285    1.25000    8.12500
286    1.25000    8.75000
287    0.00000    9.37500
288    1.25000    9.37500
289    1.25000   10.00000
  1         11       0.000000  0.000000  0.000000
  2          1       0.000000  0.000000  0.000000
  3          1       0.000000  0.000000  0.000000
  4          1       0.000000  0.000000  0.000000
  5          1       0.000000  0.000000  0.000000
  6          1       0.000000  0.000000  0.000000
  7          1       0.000000  0.000000  0.000000
  8          1       0.000000  0.000000  0.000000
  9          1       0.000000  0.000000  0.000000
 10         11       0.000000  0.000000  0.000000
 18          1       0.000000  0.000000  0.000000
 19         11       0.000000  0.000000  0.000000
 27          1       0.000000  0.000000  0.000000
 28         11       0.000000  0.000000  0.000000
 36          1       0.000000  0.000000  0.000000
 37         11       0.000000  0.000000  0.000000
 45          1       0.000000  0.000000  0.000000
 46         11       0.000000  0.000000  0.000000
 54          1       0.000000  0.000000  0.000000
 55         11       0.000000  0.000000  0.000000
 63          1       0.000000  0.000000  0.000000
 64         11       0.000000  0.000000  0.000000
 72          1       0.000000  0.000000  0.000000
 73         11       0.000000  0.000000  0.000000
 74         11       0.437500  0.000000  0.000000
```

75	11	0.750000	0.000000	0.000000
76	11	0.937500	0.000000	0.000000
77	11	1.000000	0.000000	0.000000
78	11	0.937500	0.000000	0.000000
79	11	0.750000	0.000000	0.000000
80	11	0.437500	0.000000	0.000000
81	1	0.000000	0.000000	0.000000
82	11	0.000000	0.000000	0.000000
86	1	0.000000	0.000000	0.000000
90	1	0.000000	0.000000	0.000000
94	1	0.000000	0.000000	0.000000
98	1	0.000000	0.000000	0.000000
102	1	0.000000	0.000000	0.000000
106	1	0.000000	0.000000	0.000000
110	1	0.000000	0.000000	0.000000
113	1	0.000000	0.000000	0.000000
114	1	0.000000	0.000000	0.000000
115	11	0.000000	0.000000	0.000000
139	1	0.000000	0.000000	0.000000
140	11	0.000000	0.000000	0.000000
164	1	0.000000	0.000000	0.000000
165	11	0.000000	0.000000	0.000000
189	1	0.000000	0.000000	0.000000
190	11	0.000000	0.000000	0.000000
214	1	0.000000	0.000000	0.000000
215	11	0.000000	0.000000	0.000000
239	1	0.000000	0.000000	0.000000
240	11	0.000000	0.000000	0.000000
264	1	0.000000	0.000000	0.000000
265	11	0.000000	0.000000	0.000000
266	11	0.234375	0.000000	0.000000
269	11	0.609375	0.000000	0.000000
272	11	0.859375	0.000000	0.000000
275	11	0.984375	0.000000	0.000000
278	11	0.984375	0.000000	0.000000
281	11	0.859375	0.000000	0.000000
284	11	0.609375	0.000000	0.000000
287	11	0.234375	0.000000	0.000000
289	1	0.000000	0.000000	0.000000
1				

```
0.200E+02  0.0   1.0    0.6  1.0  0.0  0.100E+08 1.0
0.650E+00 0.100E+01
                16           3           4           5   1
                32           3           4           5   1
                48           3           4           5   1
                64           3           4           5   1
                80           3           4           5   1
                96           3           4           5   1
               112           3           4           5   1
               128           3           4           5   1
LOADING
     0     0     0     0
    1.00000    0.00100    49     1     1
```

Output Data File

The results from output file EPOUT.RES are listed as follow:

```
Appendix A example 1: friction f=0.65,p=0.6,parabolic inlet


NPOIN = 289  NELEM = 128  NVFIX = 64   NTYPE = 2  NNODE = 6

NMATS =   1  NGAUS =   7  NEVAB = 12   NNODP = 3

NBELM =   8  NCONV =   0  NSTRE =  3   NLINR = 1  NPOWR = 1
```

ELEMENT	PROPERTY	NODE NUMBERS					
1	1	1	84	11	83	10	82
2	1	1	86	2	85	11	84
3	1	2	88	12	87	11	85
4	1	2	90	3	89	12	88
5	1	3	92	13	91	12	89
6	1	3	94	4	93	13	92
7	1	4	96	14	95	13	93
8	1	4	98	5	97	14	96
9	1	5	100	15	99	14	97
10	1	5	102	6	101	15	100

```
. . .
. . .
```

116	1	65	244	66	271	75	270
117	1	66	273	76	272	75	271
118	1	66	247	67	274	76	273
119	1	67	276	77	275	76	274
120	1	67	250	68	277	77	276
121	1	68	279	78	278	77	277
122	1	68	253	69	280	78	279
123	1	69	282	79	281	78	280
124	1	69	256	70	283	79	282
125	1	70	285	80	284	79	283
126	1	70	259	71	286	80	285
127	1	71	288	81	287	80	286
128	1	71	262	72	289	81	288

NODE	X	Y
1	40.000	0.000
2	40.000	1.250
3	40.000	2.500
4	40.000	3.750
5	40.000	5.000
6	40.000	6.250
7	40.000	7.500
8	40.000	8.750
9	40.000	10.000
10	30.000	0.000
11	30.000	1.250
12	30.000	2.500
13	30.000	3.750
14	30.000	5.000
15	30.000	6.250
16	30.000	7.500
17	30.000	8.750
18	30.000	10.000
19	20.000	0.000
20	20.000	1.250
21	20.000	2.500

22	20.000	3.750
23	20.000	5.000
24	20.000	6.250
25	20.000	7.500
26	20.000	8.750
27	20.000	10.000
28	15.000	0.000
29	15.000	1.250
30	15.000	2.500
31	15.000	3.750
32	15.000	5.000
33	15.000	6.250
34	15.000	7.500
35	15.000	8.750
36	15.000	10.000
37	10.000	0.000
38	10.000	1.250
39	10.000	2.500
...		
...		
268	1.250	1.250
269	0.000	1.875
270	1.250	1.875
271	1.250	2.500
272	0.000	3.125
273	1.250	3.125
274	1.250	3.750
275	0.000	4.375
276	1.250	4.375
277	1.250	5.000
278	0.000	5.625
279	1.250	5.625
280	1.250	6.250
281	0.000	6.875
282	1.250	6.875
283	1.250	7.500
284	0.000	8.125
285	1.250	8.125

286	1.250	8.750	
287	0.000	9.375	
288	1.250	9.375	
289	1.250	10.000	

NODE	CODE	FIXED VALUES		
1	11	0.000000	0.000000	0.000000
2	1	0.000000	0.000000	0.000000
3	1	0.000000	0.000000	0.000000
4	1	0.000000	0.000000	0.000000
5	1	0.000000	0.000000	0.000000
6	1	0.000000	0.000000	0.000000
7	1	0.000000	0.000000	0.000000
8	1	0.000000	0.000000	0.000000
9	1	0.000000	0.000000	0.000000
10	11	0.000000	0.000000	0.000000
18	1	0.000000	0.000000	0.000000
19	11	0.000000	0.000000	0.000000
27	1	0.000000	0.000000	0.000000
28	11	0.000000	0.000000	0.000000
36	1	0.000000	0.000000	0.000000
37	11	0.000000	0.000000	0.000000
45	1	0.000000	0.000000	0.000000
46	11	0.000000	0.000000	0.000000
54	1	0.000000	0.000000	0.000000
55	11	0.000000	0.000000	0.000000
63	1	0.000000	0.000000	0.000000
64	11	0.000000	0.000000	0.000000
72	1	0.000000	0.000000	0.000000
73	11	0.000000	0.000000	0.000000
74	11	0.437500	0.000000	0.000000
75	11	0.750000	0.000000	0.000000
76	11	0.937500	0.000000	0.000000
77	11	1.000000	0.000000	0.000000
78	11	0.937500	0.000000	0.000000
79	11	0.750000	0.000000	0.000000
80	11	0.437500	0.000000	0.000000
81	1	0.000000	0.000000	0.000000
82	11	0.000000	0.000000	0.000000

86	1	0.000000	0.000000	0.000000
90	1	0.000000	0.000000	0.000000
94	1	0.000000	0.000000	0.000000
98	1	0.000000	0.000000	0.000000
102	1	0.000000	0.000000	0.000000
106	1	0.000000	0.000000	0.000000
110	1	0.000000	0.000000	0.000000
113	1	0.000000	0.000000	0.000000
114	1	0.000000	0.000000	0.000000
115	11	0.000000	0.000000	0.000000
139	1	0.000000	0.000000	0.000000
140	11	0.000000	0.000000	0.000000
164	1	0.000000	0.000000	0.000000
165	11	0.000000	0.000000	0.000000
189	1	0.000000	0.000000	0.000000
190	11	0.000000	0.000000	0.000000
214	1	0.000000	0.000000	0.000000
215	11	0.000000	0.000000	0.000000
239	1	0.000000	0.000000	0.000000
240	11	0.000000	0.000000	0.000000
264	1	0.000000	0.000000	0.000000
265	11	0.000000	0.000000	0.000000
266	11	0.234375	0.000000	0.000000
269	11	0.609375	0.000000	0.000000
272	11	0.859375	0.000000	0.000000
275	11	0.984375	0.000000	0.000000
278	11	0.984375	0.000000	0.000000
281	11	0.859375	0.000000	0.000000
284	11	0.609375	0.000000	0.000000
287	11	0.234375	0.000000	0.000000
289	1	0.000000	0.000000	0.000000

```
NUMBER      ELEMENT PROPERTIES
1      0.2000E+02  0.0000E+00  0.1000E+01  0.6000E+00
       0.1000E+01  0.0000E+00  0.1000E+08  0.1000E+01

FRICTION COEEF. =            0.650000000000
RAM SPEED =          1.00000000000
```

INFORMATION FOR FRICTIONAL ELEMENTS

```
 16    3    4    5    1
 32    3    4    5    1
 48    3    4    5    1
 64    3    4    5    1
 80    3    4    5    1
 96    3    4    5    1
112    3    4    5    1
128    3    4    5    1
```

MAXIMUM FRONTWIDTH ENCOUNTERED =   44 MFRON= 1999
O LOADING
      0   0   0
O   TOTAL NODAL FORCES FOR EACH ELEMENT

(There is no nodal force given. The program printed
 all zero values here which are omitted for the clarity)

O   DUMY FACTOR = 1.00000 CONVERGENCE TOLERANCE = 0.00100
MAX. NO. OF ITERATIONS =   49

INITIAL OUTPUT PARAMETER = 1   FINAL OUTPUT PARAMETER = 1
ITRATION NO.   1 MAXIMUM CHANGE AT   153
O   CONVERGENCE CODE = 1 NORM OF VELOCITY SUMRATIO=0.10E+03
MAXIMUM CHANGE =   0.100000E+01

TOTAL FORCES ON THE RAM WITH PRESCRIBED VELOCITY

  U-FORCE =   6904.18867357
  ITRATION NO.   2 MAXIMUM CHANGE AT   203
O CONVERGENCE CODE=1 NORM OF VELOCITY SUMRATIO=0.35552E+01
MAXIMUM CHANGE =   0.540642E-01

TOTAL FORCES ON THE RAM WITH PRESCRIBED VELOCITY

  U-FORCE =   57.1014635098

```
    ·
    ·
 ITRATION NO.           49 MAXIMUM CHANGE AT          107
0 CONVERGENCE CODE=1 NORM OF VELOCITY SUMRATIO=0.29072E-01
MAXIMUM CHANGE =   0.883381E-03


 TOTAL FORCES ON THE RAM WITH PRESCRIBED VELOCITY


   U-FORCE =           56.7027277296
0     VELOCITIES
0     NODE     X-VELO.         Y-VELO.
        1  0.000000E+00  0.000000E+00
        2  0.400685E+00  0.000000E+00
        3  0.655078E+00  0.000000E+00
        4  0.789581E+00  0.000000E+00
        5  0.837750E+00  0.000000E+00
        6  0.839951E+00  0.000000E+00
        7  0.791470E+00  0.000000E+00
        8  0.664623E+00  0.000000E+00
        9  0.433572E+00  0.000000E+00
       10  0.000000E+00  0.000000E+00
       11  0.432442E+00  0.762712E-04
       12  0.693209E+00  0.571974E-02
       13  0.822053E+00  0.127513E-01
       14  0.862593E+00  0.187458E-01
       15  0.857085E+00  0.226919E-01
       16  0.793030E+00  0.233234E-01
       17  0.627999E+00  0.170243E-01
       18  0.324278E+00  0.000000E+00
       19  0.000000E+00  0.000000E+00
       20  0.484494E+00  0.833894E-03
       21  0.759426E+00  0.847976E-02
       22  0.880482E+00  0.162311E-01
       23  0.905824E+00  0.222889E-01
       24  0.888412E+00  0.268238E-01
       25  0.790372E+00  0.290475E-01
       26  0.540022E+00  0.235722E-01
       27  0.640710E-01  0.000000E+00
       28  0.000000E+00  0.000000E+00
       29  0.482626E+00  0.168378E-02
```

```
30   0.763083E+00   0.757696E-03
31   0.886044E+00   0.502313E-03
32   0.911905E+00   0.480331E-03
33   0.889363E+00   0.923826E-03
34   0.774186E+00  -0.162986E-03
35   0.502951E+00  -0.361428E-02
36   0.266664E-01   0.000000E+00
37   0.000000E+00   0.000000E+00
38   0.487328E+00  -0.281579E-02
39   0.766826E+00  -0.496492E-02
40   0.890215E+00  -0.476907E-02
41   0.915941E+00  -0.327457E-02
42   0.893522E+00  -0.118676E-02
43   0.777592E+00   0.814371E-03
44   0.508533E+00   0.216220E-02
45   0.217848E-01   0.000000E+00
46   0.000000E+00   0.000000E+00
47   0.481172E+00  -0.173905E-02
48   0.765001E+00  -0.399815E-02
49   0.892394E+00  -0.429509E-02
50   0.919415E+00  -0.254924E-02
51   0.893698E+00   0.538260E-03
52   0.770318E+00   0.115885E-02
53   0.497631E+00  -0.150574E-02
54   0.344874E-01   0.000000E+00
55   0.000000E+00   0.000000E+00
56   0.476073E+00  -0.126347E-02
57   0.764067E+00  -0.376716E-02
58   0.898218E+00  -0.346254E-02
59   0.929129E+00   0.324107E-02
60   0.898259E+00   0.129435E-01
61   0.763160E+00   0.145236E-01
62   0.487070E+00   0.683551E-02
63   0.525347E-01   0.000000E+00
64   0.000000E+00   0.000000E+00
65   0.458590E+00  -0.105459E-01
66   0.758841E+00  -0.224454E-01
67   0.919583E+00  -0.144076E-01
68   0.961254E+00   0.168719E-01
69   0.910415E+00   0.437092E-01
```

70	0.749220E+00	0.450391E-01
71	0.458332E+00	0.265038E-01
72	0.529229E-01	0.000000E+00
73	0.000000E+00	0.000000E+00
74	0.437500E+00	0.000000E+00
75	0.750000E+00	0.000000E+00
76	0.937500E+00	0.000000E+00
77	0.100000E+01	0.000000E+00
78	0.937500E+00	0.000000E+00
79	0.750000E+00	0.000000E+00
80	0.437500E+00	0.000000E+00
81	-0.106310E-01	0.000000E+00
82	0.000000E+00	0.000000E+00
83	0.245582E+00	0.189447E-04
84	0.245275E+00	0.192383E-04
85	0.408626E+00	0.200404E-02
86	0.237336E+00	0.000000E+00
87	0.582611E+00	0.347169E-02
88	0.567974E+00	0.341518E-02
89	0.664612E+00	0.579835E-02
90	0.558441E+00	0.000000E+00
91	0.772081E+00	0.986059E-02
92	0.753486E+00	0.755645E-02
93	0.797699E+00	0.958597E-02
94	0.745368E+00	0.000000E+00
95	0.849091E+00	0.162793E-01
96	0.833229E+00	0.110848E-01
97	0.843961E+00	0.126375E-01
98	0.827018E+00	0.000000E+00
99	0.862875E+00	0.212850E-01
100	0.853301E+00	0.136241E-01
101	0.844235E+00	0.146949E-01
102	0.849017E+00	0.000000E+00
103	0.834452E+00	0.239205E-01
104	0.836208E+00	0.148529E-01
105	0.791861E+00	0.149504E-01
106	0.835818E+00	0.000000E+00
107	0.725433E+00	0.218460E-01
108	0.751017E+00	0.133758E-01
109	0.655468E+00	0.110867E-01

110	0.760174E+00	0.000000E+00
111	0.499897E+00	0.104790E-01
112	0.558836E+00	0.683085E-02
113	0.406249E+00	0.000000E+00
114	0.586160E+00	0.000000E+00
115	0.000000E+00	0.000000E+00
116	0.261932E+00	0.208354E-03
117	0.258594E+00	0.208623E-03
118	0.445762E+00	0.344266E-02
119	0.632921E+00	0.596011E-02
120	0.615042E+00	0.539262E-02
121	0.716461E+00	0.101049E-01
122	0.826903E+00	0.137188E-01
123	0.805753E+00	0.128801E-01
124	0.845723E+00	0.171570E-01
125	0.894821E+00	0.202644E-01
126	0.876134E+00	0.198043E-01
127	0.881145E+00	0.230370E-01
128	0.898204E+00	0.251566E-01
129	0.883128E+00	0.251387E-01
130	0.868280E+00	0.276628E-01
131	0.846781E+00	0.289172E-01
132	0.842354E+00	0.286392E-01
133	0.786326E+00	0.293859E-01
134	0.676943E+00	0.282674E-01
135	0.704218E+00	0.272622E-01
136	0.580030E+00	0.234358E-01
137	0.318265E+00	0.154273E-01
138	0.416532E+00	0.142961E-01
139	0.209444E+00	0.000000E+00
140	0.000000E+00	0.000000E+00
141	0.264834E+00	0.420845E-03
142	0.261465E+00	0.421000E-03
143	0.485695E+00	-0.210749E-04
144	0.646979E+00	0.801584E-03
145	0.646695E+00	0.164580E-03
146	0.762773E+00	0.192498E-02
147	0.842507E+00	0.542691E-03
148	0.841499E+00	0.246941E-02
149	0.884170E+00	0.452804E-02

150	0.907281E+00	0.617543E-03
151	0.906417E+00	0.509706E-02
152	0.909384E+00	0.686165E-02
153	0.906784E+00	0.841327E-03
154	0.905377E+00	0.748257E-02
155	0.890784E+00	0.820160E-02
156	0.848608E+00	0.237374E-03
157	0.848884E+00	0.846348E-02
158	0.784623E+00	0.742187E-02
159	0.665473E+00	-0.262340E-02
160	0.670031E+00	0.613345E-02
161	0.517449E+00	0.345711E-02
162	0.303099E+00	-0.401837E-02
163	0.299907E+00	0.103446E-02
164	0.165833E-01	0.000000E+00
165	0.000000E+00	0.000000E+00
166	0.260378E+00	-0.704067E-03
167	0.266009E+00	-0.703907E-03
168	0.487170E+00	-0.537326E-03
169	0.643208E+00	-0.360756E-02
170	0.645313E+00	-0.232505E-03
171	0.764770E+00	-0.183822E-03
172	0.843153E+00	-0.463585E-02
173	0.842946E+00	0.528612E-04
174	0.887181E+00	0.486913E-03
175	0.910813E+00	-0.389761E-02
176	0.908773E+00	0.884056E-03
177	0.912387E+00	0.151472E-02
178	0.911465E+00	-0.211649E-02
179	0.908825E+00	0.190498E-02
180	0.890289E+00	0.256412E-02
181	0.850650E+00	-0.138265E-03
182	0.847800E+00	0.303609E-02
183	0.774524E+00	0.331873E-02
184	0.661809E+00	0.170557E-02
185	0.660478E+00	0.352712E-02
186	0.503835E+00	0.349001E-02
187	0.290417E+00	0.244722E-02
188	0.296648E+00	0.282154E-02
189	0.379057E-01	0.000000E+00

```
190   0.000000E+00   0.000000E+00
191   0.257101E+00  -0.434865E-03
192   0.258839E+00  -0.434775E-03
193   0.482973E+00  -0.127006E-02
194   0.638708E+00  -0.243522E-02
195   0.642243E+00  -0.324255E-02
196   0.764730E+00  -0.341603E-02
197   0.842778E+00  -0.370862E-02
198   0.844258E+00  -0.431030E-02
199   0.890395E+00  -0.353450E-02
200   0.913515E+00  -0.313182E-02
201   0.912678E+00  -0.328019E-02
202   0.917308E+00  -0.228733E-02
203   0.915499E+00  -0.115814E-02
204   0.912781E+00  -0.126604E-02
205   0.894382E+00  -0.110794E-02
206   0.851321E+00   0.329535E-03
207   0.849928E+00  -0.544991E-03
208   0.776679E+00  -0.231567E-02
209   0.660251E+00  -0.177830E-02
210   0.660192E+00  -0.252942E-02
211   0.506047E+00  -0.503783E-02
212   0.299716E+00  -0.397473E-02
213   0.295244E+00  -0.454166E-02
214   0.211473E-01   0.000000E+00
215   0.000000E+00   0.000000E+00
216   0.254563E+00  -0.315989E-03
217   0.255826E+00  -0.315886E-03
218   0.478158E+00  -0.115592E-02
219   0.635244E+00  -0.229473E-02
220   0.638212E+00  -0.265119E-02
221   0.762772E+00  -0.245153E-02
222   0.843504E+00  -0.292662E-02
223   0.844961E+00  -0.337348E-02
224   0.892826E+00  -0.147945E-02
225   0.921620E+00   0.181606E-03
226   0.917397E+00  -0.315760E-03
227   0.922136E+00   0.240302E-02
228   0.925444E+00   0.794899E-02
229   0.917878E+00   0.497476E-02
```

230	0.896687E+00	0.596227E-02
231	0.851093E+00	0.135541E-01
232	0.848804E+00	0.728184E-02
233	0.769877E+00	0.497330E-02
234	0.648519E+00	0.101378E-01
235	0.653069E+00	0.372550E-02
236	0.496868E+00	-0.519725E-03
237	0.296220E+00	0.154749E-02
238	0.298537E+00	-0.128982E-02
239	0.461960E-01	0.000000E+00
240	0.000000E+00	0.000000E+00
241	0.239647E+00	-0.263661E-02
242	0.250193E+00	-0.263648E-02
243	0.470439E+00	-0.669165E-02
244	0.624336E+00	-0.172601E-01
245	0.633127E+00	-0.102981E-01
246	0.761067E+00	-0.113841E-01
247	0.858026E+00	-0.208961E-01
248	0.850374E+00	-0.102210E-01
249	0.902335E+00	-0.419161E-02
250	0.958302E+00	-0.634450E-03
251	0.933588E+00	0.301621E-02
252	0.939928E+00	0.113397E-01
253	0.954622E+00	0.302288E-01
254	0.933048E+00	0.194326E-01
255	0.906436E+00	0.223283E-01
256	0.851468E+00	0.445118E-01
257	0.848039E+00	0.252297E-01
258	0.760824E+00	0.215652E-01
259	0.622507E+00	0.363457E-01
260	0.636407E+00	0.175058E-01
261	0.477125E+00	0.977950E-02
262	0.267080E+00	0.140734E-01
263	0.289159E+00	0.177333E-02
264	0.529547E-01	0.000000E+00
265	0.000000E+00	0.000000E+00
266	0.234375E+00	0.000000E+00
267	0.234375E+00	-0.695806E-07
268	0.442772E+00	-0.263642E-02
269	0.609375E+00	0.000000E+00

```
270   0.614647E+00  -0.790908E-02
271   0.752210E+00  -0.108839E-01
272   0.859375E+00   0.000000E+00
273   0.861585E+00  -0.130942E-01
274   0.933021E+00  -0.110847E-01
275   0.984375E+00   0.000000E+00
276   0.979896E+00  -0.660557E-02
277   0.990313E+00   0.121433E-02
278   0.984375E+00   0.000000E+00
279   0.974688E+00   0.109009E-01
280   0.930729E+00   0.176102E-01
281   0.859375E+00   0.000000E+00
282   0.852604E+00   0.243814E-01
283   0.749805E+00   0.247139E-01
284   0.609375E+00   0.000000E+00
285   0.609180E+00   0.249088E-01
286   0.442708E+00   0.202750E-01
287   0.234375E+00   0.000000E+00
288   0.239583E+00   0.150671E-01
289   0.221396E-01   0.000000E+00
```

A.6.2   Example 2: Flow over a step

The data below is the Newtonian flow case of the example used in chapter 4. The details are given in Section 4.3.3 and in Figures 4.2 and 4.3. There are a total 472 elements and 1021 nodes in the mesh.

Input Data File

Contents of the **EPIN.DAT** for this example are as follows:

```
APPENDIX A EXAMPLE 2: FLOW OVER A STEP WITH D=1, P=1,MU=1
1021  472  152    2    6    1    7    3    0    0    3    1    1
      1    1  977  958  956  957  959  961
      2    1  857  829  828  827  825  826
      3    1  891  858  857  856  859  861
      4    1  857  826  825  824  859  856
```

5	1	825	772	770	771	798	797
6	1	859	824	825	797	798	800
7	1	922	894	891	892	890	893
8	1	891	861	859	860	890	892
9	1	592	566	565	564	567	569
10	1	451	410	408	409	486	452
11	1	547	518	519	488	486	487
12	1	550	546	545	544	547	549
13	1	545	523	521	520	519	522
14	1	521	491	489	490	519	520
15	1	545	522	519	518	547	544
16	1	376	340	338	339	373	372
17	1	17	8	6	7	14	13
18	1	30	19	17	18	45	29
19	1	6	2	1	3	14	7
20	1	45	18	17	13	14	16
21	1	48	31	30	29	45	46
22	1	9	11	21	20	25	10
23	1	21	23	33	32	41	22
24	1	33	35	51	52	54	34
25	1	51	53	73	56	54	52

.
.
.

464	1	572	556	557	558	596	573
465	1	585	571	572	574	607	586
466	1	673	631	629	625	626	628
467	1	622	587	585	586	607	606
468	1	629	623	622	606	607	609
469	1	596	597	599	600	626	598
470	1	607	574	572	573	596	595
471	1	629	609	607	608	626	625
472	1	607	595	596	598	626	608

1	0.40000D+01	0.20122D+00
2	0.40000D+01	0.10061D+00
3	0.39031D+01	0.19333D+00
4	0.39200D+01	0.31269D+00
5	0.40000D+01	0.29989D+00
6	0.40000D+01	0.00000D+00
7	0.39031D+01	0.92726D-01

```
    8  0.38986D+01      0.00000D+00
    9  0.40000D+01      0.39856D+00
   10  0.39200D+01      0.41136D+00
   11  0.40000D+01      0.49712D+00
   12  0.38231D+01      0.30480D+00
   13  0.38017D+01      0.92726D-01
   14  0.38061D+01      0.18545D+00
   15  0.37156D+01      0.26631D+00
   16  0.36882D+01      0.16365D+00
   17  0.37972D+01      0.00000D+00
   18  0.36837D+01      0.70928D-01
   19  0.36977D+01      0.00000D+00
   20  0.39200D+01      0.50992D+00
   21  0.40000D+01      0.59569D+00
   22  0.39100D+01      0.63677D+00
              .
              .
              .
 1006  0.40345D+00      0.39808D-01
 1007  0.00000D+00      0.21128D+00
 1008  0.92742D-01      0.18694D+00
 1009  0.00000D+00      0.10564D+00
 1010  0.25402D+00      0.13356D+00
 1011  0.18548D+00      0.16259D+00
 1012  0.92742D-01      0.81294D-01
 1013  0.00000D+00      0.00000D+00
 1014  0.18981D+00      0.81294D-01
 1015  0.97067D-01      0.00000D+00
 1016  0.32255D+00      0.10452D+00
 1017  0.36650D+00      0.00000D+00
 1018  0.34453D+00      0.52261D-01
 1019  0.19413D+00      0.00000D+00
 1020  0.25834D+00      0.52261D-01
 1021  0.28032D+00      0.00000D+00
  828    11    0.00000E+00    0.00000E+00    0.00000E+00
  829    11    0.00000E+00    0.00000E+00    0.00000E+00
  857    11    0.00000E+00    0.00000E+00    0.00000E+00
  858    11    0.00000E+00    0.00000E+00    0.00000E+00
  891    11    0.00000E+00    0.00000E+00    0.00000E+00
  894    11    0.00000E+00    0.00000E+00    0.00000E+00
```

922	11	0.00000E+00	0.00000E+00	0.00000E+00
924	11	0.00000E+00	0.00000E+00	0.00000E+00
956	11	0.00000E+00	0.00000E+00	0.00000E+00
958	11	0.00000E+00	0.00000E+00	0.00000E+00
977	11	0.00000E+00	0.00000E+00	0.00000E+00
979	11	0.00000E+00	0.00000E+00	0.00000E+00
1001	11	0.00000E+00	0.00000E+00	0.00000E+00
1003	11	0.00000E+00	0.00000E+00	0.00000E+00
1017	11	0.00000E+00	0.00000E+00	0.00000E+00
1021	11	0.00000E+00	0.00000E+00	0.00000E+00
1019	11	0.00000E+00	0.00000E+00	0.00000E+00
1015	11	0.00000E+00	0.00000E+00	0.00000E+00
1013	11	0.00000E+00	0.00000E+00	0.00000E+00
725	11	0.00000E+00	0.00000E+00	0.00000E+00
749	11	0.00000E+00	0.00000E+00	0.00000E+00
751	11	0.00000E+00	0.00000E+00	0.00000E+00
770	11	0.00000E+00	0.00000E+00	0.00000E+00
772	11	0.00000E+00	0.00000E+00	0.00000E+00
825	11	0.00000E+00	0.00000E+00	0.00000E+00
827	11	0.00000E+00	0.00000E+00	0.00000E+00
592	11	0.00000E+00	0.00000E+00	0.00000E+00
594	11	0.00000E+00	0.00000E+00	0.00000E+00
614	11	0.00000E+00	0.00000E+00	0.00000E+00
616	11	0.00000E+00	0.00000E+00	0.00000E+00
637	11	0.00000E+00	0.00000E+00	0.00000E+00
639	11	0.00000E+00	0.00000E+00	0.00000E+00
656	11	0.00000E+00	0.00000E+00	0.00000E+00
658	11	0.00000E+00	0.00000E+00	0.00000E+00
723	11	0.00000E+00	0.00000E+00	0.00000E+00
523	11	0.00000E+00	0.00000E+00	0.00000E+00
545	11	0.00000E+00	0.00000E+00	0.00000E+00
546	11	0.00000E+00	0.00000E+00	0.00000E+00
550	11	0.00000E+00	0.00000E+00	0.00000E+00
551	11	0.00000E+00	0.00000E+00	0.00000E+00
565	11	0.00000E+00	0.00000E+00	0.00000E+00
566	11	0.00000E+00	0.00000E+00	0.00000E+00
6	11	0.00000E+00	0.00000E+00	0.00000E+00
8	11	0.00000E+00	0.00000E+00	0.00000E+00
17	11	0.00000E+00	0.00000E+00	0.00000E+00
19	11	0.00000E+00	0.00000E+00	0.00000E+00

30	11	0.00000E+00	0.00000E+00	0.00000E+00
31	11	0.00000E+00	0.00000E+00	0.00000E+00
48	11	0.00000E+00	0.00000E+00	0.00000E+00
50	11	0.00000E+00	0.00000E+00	0.00000E+00
70	11	0.00000E+00	0.00000E+00	0.00000E+00
72	11	0.00000E+00	0.00000E+00	0.00000E+00
114	11	0.00000E+00	0.00000E+00	0.00000E+00
116	11	0.00000E+00	0.00000E+00	0.00000E+00
152	11	0.00000E+00	0.00000E+00	0.00000E+00
154	11	0.00000E+00	0.00000E+00	0.00000E+00
182	11	0.00000E+00	0.00000E+00	0.00000E+00
184	11	0.00000E+00	0.00000E+00	0.00000E+00
209	11	0.00000E+00	0.00000E+00	0.00000E+00
211	11	0.00000E+00	0.00000E+00	0.00000E+00
243	11	0.00000E+00	0.00000E+00	0.00000E+00
245	11	0.00000E+00	0.00000E+00	0.00000E+00
273	11	0.00000E+00	0.00000E+00	0.00000E+00
275	11	0.00000E+00	0.00000E+00	0.00000E+00
307	11	0.00000E+00	0.00000E+00	0.00000E+00
309	11	0.00000E+00	0.00000E+00	0.00000E+00
338	11	0.00000E+00	0.00000E+00	0.00000E+00
340	11	0.00000E+00	0.00000E+00	0.00000E+00
376	11	0.00000E+00	0.00000E+00	0.00000E+00
378	11	0.00000E+00	0.00000E+00	0.00000E+00
408	11	0.00000E+00	0.00000E+00	0.00000E+00
410	11	0.00000E+00	0.00000E+00	0.00000E+00
451	11	0.00000E+00	0.00000E+00	0.00000E+00
453	11	0.00000E+00	0.00000E+00	0.00000E+00
489	11	0.00000E+00	0.00000E+00	0.00000E+00
491	11	0.00000E+00	0.00000E+00	0.00000E+00
521	11	0.00000E+00	0.00000E+00	0.00000E+00
35	1	0.00000E+00	0.00000E+00	0.00000E+00
33	1	0.00000E+00	0.00000E+00	0.00000E+00
23	1	0.00000E+00	0.00000E+00	0.00000E+00
21	1	0.00000E+00	0.00000E+00	0.00000E+00
11	1	0.00000E+00	0.00000E+00	0.00000E+00
9	1	0.00000E+00	0.00000E+00	0.00000E+00
5	1	0.00000E+00	0.00000E+00	0.00000E+00
1	1	0.00000E+00	0.00000E+00	0.00000E+00
2	1	0.00000E+00	0.00000E+00	0.00000E+00

903	11	0.00000E+00	0.00000E+00	0.00000E+00
872	11	0.00000E+00	0.00000E+00	0.00000E+00
870	11	0.00000E+00	0.00000E+00	0.00000E+00
834	11	0.00000E+00	0.00000E+00	0.00000E+00
832	11	0.00000E+00	0.00000E+00	0.00000E+00
804	11	0.00000E+00	0.00000E+00	0.00000E+00
802	11	0.00000E+00	0.00000E+00	0.00000E+00
759	11	0.00000E+00	0.00000E+00	0.00000E+00
757	11	0.00000E+00	0.00000E+00	0.00000E+00
737	11	0.00000E+00	0.00000E+00	0.00000E+00
735	11	0.00000E+00	0.00000E+00	0.00000E+00
711	11	0.00000E+00	0.00000E+00	0.00000E+00
709	11	0.00000E+00	0.00000E+00	0.00000E+00
691	11	0.00000E+00	0.00000E+00	0.00000E+00
689	11	0.00000E+00	0.00000E+00	0.00000E+00
665	11	0.00000E+00	0.00000E+00	0.00000E+00
663	11	0.00000E+00	0.00000E+00	0.00000E+00
643	11	0.00000E+00	0.00000E+00	0.00000E+00
641	11	0.00000E+00	0.00000E+00	0.00000E+00
620	11	0.00000E+00	0.00000E+00	0.00000E+00
618	11	0.00000E+00	0.00000E+00	0.00000E+00
601	11	0.00000E+00	0.00000E+00	0.00000E+00
599	11	0.00000E+00	0.00000E+00	0.00000E+00
577	11	0.00000E+00	0.00000E+00	0.00000E+00
575	11	0.00000E+00	0.00000E+00	0.00000E+00
559	11	0.00000E+00	0.00000E+00	0.00000E+00
557	11	0.00000E+00	0.00000E+00	0.00000E+00
534	11	0.00000E+00	0.00000E+00	0.00000E+00
532	11	0.00000E+00	0.00000E+00	0.00000E+00
498	11	0.00000E+00	0.00000E+00	0.00000E+00
496	11	0.00000E+00	0.00000E+00	0.00000E+00
457	11	0.00000E+00	0.00000E+00	0.00000E+00
455	11	0.00000E+00	0.00000E+00	0.00000E+00
417	11	0.00000E+00	0.00000E+00	0.00000E+00
415	11	0.00000E+00	0.00000E+00	0.00000E+00
382	11	0.00000E+00	0.00000E+00	0.00000E+00
380	11	0.00000E+00	0.00000E+00	0.00000E+00
347	11	0.00000E+00	0.00000E+00	0.00000E+00
345	11	0.00000E+00	0.00000E+00	0.00000E+00
320	11	0.00000E+00	0.00000E+00	0.00000E+00

```
 318   11   0.00000E+00   0.00000E+00   0.00000E+00
 286   11   0.00000E+00   0.00000E+00   0.00000E+00
 284   11   0.00000E+00   0.00000E+00   0.00000E+00
 255   11   0.00000E+00   0.00000E+00   0.00000E+00
 253   11   0.00000E+00   0.00000E+00   0.00000E+00
 187   11   0.00000E+00   0.00000E+00   0.00000E+00
 185   11   0.00000E+00   0.00000E+00   0.00000E+00
 157   11   0.00000E+00   0.00000E+00   0.00000E+00
 155   11   0.00000E+00   0.00000E+00   0.00000E+00
 119   11   0.00000E+00   0.00000E+00   0.00000E+00
 117   11   0.00000E+00   0.00000E+00   0.00000E+00
  89   11   0.00000E+00   0.00000E+00   0.00000E+00
  88   11   0.00000E+00   0.00000E+00   0.00000E+00
  75   11   0.00000E+00   0.00000E+00   0.00000E+00
  73   11   0.00000E+00   0.00000E+00   0.00000E+00
  53   11   0.00000E+00   0.00000E+00   0.00000E+00
  51   11   0.00000E+00   0.00000E+00   0.00000E+00
1009   11   0.10000E+01   0.00000E+00   0.00000E+00
1007   11   0.10000E+01   0.00000E+00   0.00000E+00
 985   11   0.10000E+01   0.00000E+00   0.00000E+00
 983   11   0.10000E+01   0.00000E+00   0.00000E+00
 964   11   0.10000E+01   0.00000E+00   0.00000E+00
 963   11   0.10000E+01   0.00000E+00   0.00000E+00
 937   11   0.10000E+01   0.00000E+00   0.00000E+00
 935   11   0.10000E+01   0.00000E+00   0.00000E+00
 905   11   0.10000E+01   0.00000E+00   0.00000E+00
   1
     0.20000E+02    0.00000E+00    0.10000E+01    0.10000E+01
     1.0    0.00000E+00    0.10000E+08   1.0
   0.0      0.0
POINT  GRAVITY  EDGE  LOADINGS
   0       0       0
  1.0   0.0001    50    1    1
```

Output Data File

The output data file (EPOUT.RES) for this example is listed below:

```
APPENDIX A EXAMPLE 2: FLOW OVER A STEP WITH D=1, P=1,MU=1
```

NPOIN =1021   NELEM = 472   NVFIX = 152   NTYPE = 2   NNODE = 6

NMATS =   1   NGAUS =   7   NEVAB =   12   NNODP = 3

NBELM =   0   NCONV =   0   NSTRE =   3   NLINR = 1   NPOWR = 1

ELEMENT	PROPERTY	NODE NUMBERS					
1	1	977	958	956	957	959	961
2	1	857	829	828	827	825	826
3	1	891	858	857	856	859	861
4	1	857	826	825	824	859	856
5	1	825	772	770	771	798	797
6	1	859	824	825	797	798	800
7	1	922	894	891	892	890	893
8	1	891	861	859	860	890	892
9	1	592	566	565	564	567	569
10	1	451	410	408	409	486	452
11	1	547	518	519	488	486	487
12	1	550	546	545	544	547	549
.							
.							
.							
471	1	629	609	607	608	626	625
472	1	607	595	596	598	626	608

NODE	X	Y
1	4.000	0.201
2	4.000	0.101
3	3.903	0.193
4	3.920	0.313
5	4.000	0.300
6	4.000	0.000
7	3.903	0.093
.		
.		
.		

```
1014    0.190    0.081
1015    0.097    0.000
1016    0.323    0.105
1017    0.366    0.000
1018    0.345    0.052
1019    0.194    0.000
1020    0.258    0.052
1021    0.280    0.000
```

NODE	CODE	FIXED VALUES		
828	11	0.000000	0.000000	0.000000
829	11	0.000000	0.000000	0.000000
857	11	0.000000	0.000000	0.000000
858	11	0.000000	0.000000	0.000000
891	11	0.000000	0.000000	0.000000
894	11	0.000000	0.000000	0.000000
922	11	0.000000	0.000000	0.000000
.				
.				
.				
963	11	1.000000	0.000000	0.000000
937	11	1.000000	0.000000	0.000000
935	11	1.000000	0.000000	0.000000
905	11	1.000000	0.000000	0.000000

```
NO.   ELEMENT PROPERTIES
1   0.200000E+02  0.000000E+00  0.100000E+01  0.100000E+01
    0.100000E+01  0.000000E+00  0.100000E+08  0.100000E+01

FRICTION COEEF. =         0.000000000000E+0000
RAM SPEED =       0.000000000000E+0000



MAXIMUM FRONTWIDTH ENCOUNTERED =  442 MFRON= 1999
O POINT  GRAVITY  EDGE  LOADINGS
     0    0    0
O       TOTAL NODAL FORCES FOR EACH ELEMENT
```

(There is no nodal force given. The program printed
all zero values here which are omitted for the clarity)

0 DUMY FACTOR = 1.00000 CONVERGENCE TOLERANCE = 0.00010
MAX. NO. OF ITERATIONS =   50

INITIAL OUTPUT PARAMETER = 1 FINAL OUTPUT PARAMETER=1
ITRATION NO.   1 MAXIMUM CHANGE AT   1339
0 CONVERGENCE CODE=1 NORM OF VELOCITY SUMRATIO= 0.10E+03
MAXIMUM CHANGE = 0.232354E+01

TOTAL FORCES ON THE RAM WITH PRESCRIBED VELOCITY

 U-FORCE =   18.5949089785
ITRATION NO.   2 MAXIMUM CHANGE AT -2139062144
0 CONVERGENCE CODE=0 NORM OF VELOCITY SUMRATIO= 0.00E+00
MAXIMUM CHANGE = 0.000000E+00

TOTAL FORCES ON THE RAM WITH PRESCRIBED VELOCITY

 U-FORCE =   18.5949089785
0   VELOCITIES
0   NODE   X-VELO.   Y-VELO.
            1 0.897576E+00 0.000000E+00
            2 0.505280E+00 0.000000E+00
            3 0.870910E+00 0.689209E-04
            4 0.120014E+01 0.880775E-04
            5 0.117245E+01 0.000000E+00
            6 0.000000E+00 0.000000E+00
            7 0.469784E+00 0.306285E-04
            8 0.000000E+00 0.000000E+00
            9 0.133854E+01 0.000000E+00
           10 0.135212E+01 0.104059E-03
           11 0.139588E+01 0.000000E+00
           12 0.118334E+01 0.202136E-03
           13 0.469851E+00 0.306180E-04
           14 0.843626E+00 0.122513E-03
           15 0.109126E+01 0.319122E-03
           16 0.764541E+00 0.159451E-03

```
17   0.000000E+00   0.000000E+00
18   0.368111E+00   0.595424E-04
19   0.000000E+00   0.000000E+00
20   0.139538E+01   0.118222E-03
21   0.134478E+01   0.000000E+00
22   0.129142E+01   0.105809E-03

        .

        .

        .

939   0.130051E+01   0.178055E+00
940   0.125535E+01   0.164938E+00
941   0.128467E+01   0.140844E+00
942   0.132996E+01   0.952556E-01
943   0.132258E+01   0.653097E-01
944   0.128989E+01   0.968851E-01
945   0.128798E+01   0.389585E-02
946   0.119404E+01  -0.499632E-01
947   0.127438E+01   0.644936E-01
948   0.937275E+00   0.156700E+00
949   0.112594E+01   0.179945E+00
950   0.115979E+01   0.178261E+00
951   0.993019E+00   0.145506E+00
952   0.536285E+00   0.736977E-01
953   0.708831E+00   0.111882E+00
954   0.794247E+00   0.963547E-01
955   0.590270E+00   0.641876E-01
956   0.000000E+00   0.000000E+00
957   0.166485E+00   0.105275E-01
958   0.000000E+00   0.000000E+00
959   0.326796E+00   0.421150E-01
960   0.374069E+00   0.301209E-01
961   0.193504E+00   0.105288E-01
962   0.102068E+01  -0.425506E-01
963   0.100000E+01   0.000000E+00
964   0.100000E+01   0.000000E+00
965   0.124944E+01   0.152037E+00
966   0.117259E+01   0.153809E+00
967   0.104201E+01   0.143358E+00
968   0.112419E+01   0.159056E+00
969   0.120378E+01   0.198334E-01
```

970	0.108274E+01	-0.119968E+00
971	0.102068E+01	0.496909E-03
972	0.110720E+01	0.561420E-01
973	0.123888E+01	0.110197E+00
974	0.120028E+01	0.166888E+00
975	0.114931E+01	0.136495E+00
976	0.655084E+00	0.545116E-01
977	0.000000E+00	0.000000E+00
978	0.215278E+00	0.806694E-02
979	0.000000E+00	0.000000E+00
980	0.413985E+00	0.322685E-01
981	0.417037E+00	0.231785E-01
982	0.238041E+00	0.806684E-02
983	0.100000E+01	0.000000E+00
984	0.101388E+01	0.735414E-01
985	0.100000E+01	0.000000E+00
986	0.101112E+01	0.137323E+00
987	0.113951E+01	0.217595E+00
988	0.109348E+01	0.155872E+00
989	0.997363E+00	0.158835E+00
990	0.990625E+00	0.196782E+00
991	0.105553E+01	0.183135E+00
992	0.101388E+01	0.106709E+00
993	0.107326E+01	0.267681E+00
994	0.652972E+00	0.711495E-01
995	0.878785E+00	0.953092E-01
996	0.877854E+00	0.125825E+00
997	0.738880E+00	0.104827E+00
998	0.874806E+00	0.150456E+00
999	0.877670E+00	0.160951E+00
1000	0.746263E+00	0.132539E+00
1001	0.000000E+00	0.000000E+00
1002	0.216421E+00	0.114939E-01
1003	0.000000E+00	0.000000E+00
1004	0.414344E+00	0.459749E-01
1005	0.503575E+00	0.476636E-01
1006	0.259979E+00	0.114926E-01
1007	0.100000E+01	0.000000E+00
1008	0.103709E+01	0.167267E+00
1009	0.100000E+01	0.000000E+00

```
1010    0.763186E+00    0.119449E+00
1011    0.917898E+00    0.198981E+00
1012    0.806628E+00    0.497444E-01
1013    0.000000E+00    0.000000E+00
1014    0.687833E+00    0.497460E-01
1015    0.000000E+00    0.000000E+00
1016    0.623252E+00    0.108350E+00
1017    0.000000E+00    0.000000E+00
1018    0.361132E+00    0.270897E-01
1019    0.000000E+00    0.000000E+00
1020    0.450482E+00    0.270871E-01
1021    0.000000E+00    0.000000E+00
```

Appendix B

Software Description for NFLOWG

B.1 Introduction

In this appendix we present a detailed description of the program **NFLOWG** with complete instructions for the user. This program allows a user to perform an analysis of Newtonian and non-Newtonian flow for 2-D plane or axisymmetric problems. **NFLOWG** is based on the research work at the Department of Computer Science of Swansea University, UK. The program uses four types of elements, *i.e.*, 3 and 4-noded linear and 6 and 8-noded quadratic elements.

B.2 Glossary of Variable Names

A brief description of the main variables and arrays and the main subroutines is listed in the following sections. In the description of logical parameters, 'T' and 'F' denote the logical values of 'True' and 'False' respectively

B.2.1 Main logical variables for the code

Variable name	Description
LBODY	Flag for body force case
LDCOPL	Flag for decoupled Taylor-Galerkin formulae
LELAST	Flag for viscoelastic model: 'T'= non-Newtonian; 'F'=Newtonian;
LFDBCK	Flag for boundary conditions feedback
LFRIZV	Flag for freezing velocity field
LFRIZT	Flag for freezing stress field
LMNITR	Flag for monitoring convergence: 'T'=printout per mass-iteration step; 'F'=printout per complete time-step
LNEWT	Flag on momentum diffusion terms: 'T'=Newtonian Model; 'F'=Generalised Newtonian Model
LNCONS	Flag on momentum convection terms: 'T'=non-conservative form; 'F'=conservative form
LPLANE	Flag for plane/axisymmetric form: 'T'=plane; 'F'=axisymmetric problem
LQUADC	Flag on momentum convection terms: 'T'=exact integration; 'F' = quadrature
LQUADP	Flag on pressure-type terms (momentum): 'T'=exact integration; 'F'=quadrature
LQUADS	Flag on momentum diffusion terms: 'T'=exact integration; 'F'=quadrature
LREZRL	Flag for non-zero Reynolds number: 'T'=non-zero Reynolds number; 'F'=zero Reynolds number
LSAVR	Flag for piecewise constant viscosity
LSCALE	Flag for nondimensionalisation: 'T'=nondimensionalisation; 'F'=dimensionalisation
LSECND	Flag for *order of projection* method: 'T'=second order (Taylor-Galerkin); 'F'=first order (+NALPHA*dPdx)
LSIMPL	Flag for Semi-implicit or Explicit Taylor-Galerkin: 'T'=Semi-implicit; 'F'=explicit
LSTGE3	Flag for inclusion of projection STAGE3: 'T'=default inclusion; 'F'=exclusion

Variable name	Description
LSTRIM	Flag for stream function calculation: 'T'=calculation performed; 'F'=no calculation
LTMIND	Flag for transient boundary conditions: 'T'=time-independent boundary conditions; 'F'=time-dependent boundary conditions

B.2.2 Main variables for the geometry and mesh data

Variable name	Description
ANGLE	Domain rotation angle (in degrees)
CHRLEN	Characteristic length
CHRVEL	Characteristic velocity
DENSTY	Density
DTIM	Time step length
EL1	Relaxation time
EZERO	cut off value for non-dimensional viscosity
ITERMX	Relative time-stepping termination limit
MBAND	Maximum halfbandwidth of pressure matrix
MCTSTP	Maximum number of time steps for costing
MDIM	Maximum number of spatial dimensions
MDOF	Maximum number of degrees of freedom per node
MMONP	Maximum number of pressure nodes for monitoring
MMONV	Maximum number of velocity nodes for monitoring
MNBN	Maximum number of boundary nodes
MNODLP	Number of pressure nodes per element
MNODLV	Number of velocity nodes per element
MNQP	Number of Gauss quadrature points
MNSID	Maximum number of boundary sides
MODEL	Fluid models : 1 - General power law model; 2 - Elastic model with constant viscosity
MQBAND	Maximum half-bandwidth for quadratic stream function

Variable name	Description
MTELS	Maximum number of elements
MTNODP	Maximum number of pressure nodes
MTNODV	Maximum number of nodes
NALPHA	Waiting factor for pressure calculations
NBDNOD	Number of boundary nodes
NBKNOD	Number of inlet nodes for feed back
NDIMN	Number of dimensions
NDTYPE	Control parameter;
	1 for piecewise constant viscosity case;
	others for zero Reynolds number
NELTYP	Element type:
	1= linear triangular elements;
	2 = linear Lagrangian elements;
	3 = quadratic triangular elements;
	4 = quadratic Lagrangian elements;
	5 = quadratic Serendipity elements
NINLET	Number of nodes on inlet
NTTDOP	Number of pressure nodes (vertex nodes)
NTTELS	Number of elements
NTTNOD	Number of nodes
NUMS	Number of boundary sides specified
NVET	Number of velocity variables per mode
PN	Power law index value
POWLAW	Power law coefficient μ
RELAX	Iterative relaxation factor for Mass iteration
THETA	Crank-Nicolson factor
TOL	Tolerance for time-stepping
TOLMAS	Tolerance for Mass iteration
VISC1	Viscosity μ_1 (cf. chapter 6.4.2)
VISC2	Viscosity μ_2 (cf. chapter 6.4.2)
VISCTY	Viscosity
YIELD	Yield stress

B.2.3   Main arrays used in the Code

Array names	Description
BCSDIF(MNBN,MNVAR)	Differences on inlet
BGDP(MNSID,MDIM)	Derivative boundary values on pressure
BVAL(MNBN,MDOF)	Boundary values
COORD(MDIM,MTNODV)	Global coordinates of nodes
ELDET(MTELS)	Element areas
INLET(32)	Node number on inlet
INNODE(21)	Node number on inlet for feed back
IXNODE(21)	Node number on outlet for feed back
NBFRE(MNBN,MDOF)	Boundary freedom number, if set then value assigned
NBSID(MNSID,MDIM3)	Global node numbers on a boundary side
NELTOP(MNOD2,MTELS)	Element nodal connectivities
NF(MTNODV,MDOF)	Global freedom number for degree of freedom j of node i
NGNODE(MNBN)	Global node number of a boundary node
NPSTIR(MNODLP,MTELS)	Steering vector over elements (pressure)
NVSTIR(MNODLV,MTELS)	Steering vector over elements (velocity)
SOL(MTNODV,MNVAR)	Solution for velocities at time 'n'
SOLL(MTNODV,MNVAR)	Solution for velocities at time 'n - 1'
SOLN(MTNODV,MNVAR)	Solution for velocities at time 'n + 1'
SOLP(MTNODP)	Solution for pressure at time 'n'
SOLPL(MTNODP)	Solution for pressure at time 'n - 1'
SOLPN(MTNODP)	Solution for pressure at time 'n + 1'
SOLS(MTNODP)	Solution for stream function at time 'n'
SS(MDIM6,MNODLV,) MTNODV)	General Newtonian diffusive matrix
SYSK(MTNODP,MBAND)	Stiffness matrix for pressure
SYSM(MTNODV)	Diagonals of mass matrix for velocities
SYSS(MTNODP,MBAND)	Stiffness matrix for stream function

B.2.4   Main subroutines

Subroutine	Description
CONFRE	Constructs freedom arrays and relations of nodes and global freedoms
EFAC	Evaluates linear shape function factors and element areas
EFUN	Setup weights, linear and quadratic shape functions at Gauss points
FIXSOL	Fixes boundary conditions and calculates solution at each fractional step

Subroutine	Description
GETBND	Calculations of semi-bandwidth for system pressure stiffness matrix and for quadratic steam-function
GETDIS	Constructs information for all boundary sides
GETRH1	Calculates time-dependent right hand side vectors for step one using present time-step solution
GETRH2	Calculates time-dependent RHS vectors for step two using provisional time-step solution from step one
GETRH3	Calculates time-dependent RHS vectors for step three, using next time-step solution from step two and provisional time-step solution from step one
INPBRY	Input boundary information
INPSCA	Inputs initial conditions
INPUTD	Inputs all data
INPXTR	Inputs relevant information for feed back
LOGOUT	Outputs logical flags/parameters
NFSET	Initialisation of freedom arrays
PRMTRST	Initialisation of parameters
PSETUP	Calculation of P at t=0
REWORK	Preparation procedure before time-stepping
TAYGAL	Solves transient equations using Taylor-Galerkin method
TOLCAL	Tolerance calculation: relative differences in solution over time step

B.3   Main Routine of NFLOWG

The main routine is listed in the following lines.

```
      PROGRAM NFLOWG                                            NFLW   2
C*******************************************************************NFLW   3
C                                                                 NFLW   4
C**** Laminar CFD program with Generalised Newtonian and          NFLW   5
C     Non-Newtonian flow based on the research work at            NFLW   6
C     Department of Computer Science of Swansea University        NFLW   7
C                                                                 NFLW   8
C                                      H.C.Huang                   NFLW   9
C*******************************************************************NFLW  10
C                                                                 NFLW  11
      IMPLICIT DOUBLE PRECISION (A-H,O-Z)                         NFLW  12
      INCLUDE 'NFLOW.FIN'                                         NFLW  13
C     ===================                                         NFLW  14
C                                                                 NFLW  15
      PARAMETER (MBAND=300,MQBAND=430,MNBN=500,MNSID=300,MTNODV=3500) NFLW 16
      PARAMETER (MTNODP=1000,MTELS=1600, NDIMN=2)                 NFLW  17
      PARAMETER (MDIM=2, MNQP=7, MDOF=3,MCTSTP=250)               NFLW  18
      PARAMETER (MNODLV= 6,MNODLP=3, MMONV=12, MMONP=12, MNVAR=MDOF-1) NFLW 19
      PARAMETER (MDIM2=2*MDIM, MDOF1=MDOF+1, MMFAC=MNODLP*MDIM)   NFLW  20
      PARAMETER (MDIM6=((MDIM-1)*(MDIM+2))/2 + 1, MNOD2=MNODLV+2) NFLW  21
```

```
      PARAMETER (MDIM3=MDIM+3,MTRE1=MDIM+1)                          NFLW  22
C                                                                   NFLW  23
      CHARACTER*20 DUMPNM,STARNM,VECTNM,PRENM,PATNM,STESNM           NFLW  24
C                                                                   NFLW  25
      LOGICAL LSCALE,LMNITR,LTMIND,LELAST,LFRIZT,LFRIZV,LDCOPL,LPLANE, NFLW  26
     &        LNEWT, LSIMPL,LREZRL,LQUADS,LINERS,LQNAG,LSTGE3,       NFLW  27
     &        LQUADP,LNCONS,LSAVR,LSECND,LSTRIM,LSTGE2,LQUADC,       NFLW  28
     &        LBODY, LFDBCK,LHALFS,LPICON,LLAST                      NFLW  29
C                                                                   NFLW  30
      DIMENSION BGDP(MNSID,MDIM2),BVAL(MNBN,MDOF1),NVSTIR(MNODLV,MTELS),NFLW  31
     &          NGNODE(MNBN),NBFRE(MNBN,MDOF1),NBSID(MNSID,MDIM3),   NFLW  32
     &          COORD(MTNODV,MDIM),RAD(MNODLP,MTELS),ELFAC(MMFAC,MTELS),NFLW  33
     &          ELDET(MTELS),NELTOP(MNOD2,MTELS),NF(MTNODV,MDOF),    NFLW  34
     &          NG(MTNODV),NH(MTNODV),NSTEER(MNODLP,MNSID),          NFLW  35
     &          NPSTIR(MNODLP,MTELS),SYSK(MTNODP,MBAND),             NFLW  36
     &          SYSS(MTNODP,MBAND),SYSM(MTNODV),SYSG(MTNODV,MNVAR),  NFLW  37
     &          SOL(MTNODV,MNVAR),SOLN(MTNODV,MNVAR),SOLP(MTNODP),   NFLW  38
     &          SOLPN(MTNODP),SOLS(MTNODP),WORK(MTNODV,MNVAR),       NFLW  39
     &          VBAR(MNVAR+1),FUNP(MNODLP,MNQP),FUNV(MNODLV,MNQP),   NFLW  40
     &          GDV(MDIM,MNODLV),WGHT(MNQP),B(MNODLV,MNVAR),         NFLW  41
     &          C(MNODLV,MNVAR),DD(MNODLP,MNODLP),                   NFLW  42
     &          MONV(MMONV),MONP(MMONP),SMATE(MNODLV,MNODLV),        NFLW  43
     &          SS(MDIM6,MNODLV,MNODLV),ALMATE(MNODLP,MNODLV,MDIM),  NFLW  44
     &          CMATE(MNODLV,MNODLV),CCMATA(MNODLV,MNODLV),          NFLW  45
     &          CCMATB(MNODLV-1,MNODLV),CCMATC(MNODLV-2,MNODLV),     NFLW  46
     &          CCMATD(MNODLV-3,MNODLV),CCMATE(MNODLV-4,MNODLV),     NFLW  47
     &          CCMATF(MNODLV-5,MNODLV),CCMATG(MNODLV-5,MNODLV),     NFLW  48
     &          CCMATH(MNODLV-5,MNODLV),CCMATI(MNODLV-5,MNODLV),     NFLW  49
     &          CCMATJ(MNODLV),AMMATD(MNODLV),AMMATE(MNODLV,MNODLV)  NFLW  50
C                                                                   NFLW  51
      DIMENSION BODYF(MDIM),NSCBRY(MMONV),BCSDIF(MNBN,MNVAR),        NFLW  52
     &          DMATE(MNODLV,MNODLV,MDIM),FMATE(MNODLV,MNODLV,MDIM), NFLW  53
     &          AMMAT1(MNODLV,MNODLV),AMMAT2(MNODLV,MNODLV),         NFLW  54
     &          AMMAT3(MNODLV,MNODLV),SYSQ(MTNODV,MQBAND),SOLQ(MTNODV), NFLW  55
     &          AMATD1(MNODLV),AMATD2(MNODLV),AMATD3(MNODLV),        NFLW  56
     &          INNODE(21),IXNODE(21),INLET(32),FLUX(MTNODV,MDIM6),  NFLW  57
     &          CTSTEP(0:MCTSTP),MONS(MMONV)                         NFLW  58
C                                                                   NFLW  59
      DIMENSION GEOM(MNODLV,MDIM+1),BT(MNODLP),CT(MNODLP),           NFLW  60
     &          RHSADD(MTNODV,MNVAR),DEL(MTNODV,MNVAR),              NFLW  61
     &          DELR(MTNODV,MNVAR)                                   NFLW  62
C                                                                   NFLW  64
C**** input data ****                                               NFLW  65
C                                                                   NFLW  66
      CALL INPUTD(LPICON,LLAST,LNEWT,LQUADS,LQNAG,LQUADP,LSECND,     NFLW  67
     &            LSCALE,LTMIND,LMNITR,LSTRIM,LSIMPL,LQUADC,LPLANE,  NFLW  68
     &            LELAST,LBODY,LREZRL,LINERS,LFDBCK,LSAVR,LSTGE2,    NFLW  69
     &            LSTGE3,LDCOPL,LFRIZV,LFRIZT,MNVAR,BODYF,MDIM,MONS,MONP,NFLW  70
     &            MONV,NSCBRY,MMONV, MMONP, CTSTEP, BGDP,DUMPNM, STARNM,NFLW  71
     &            VECTNM, PRENM, PATNM,COORD, NELTOP, MCTSTP, MNSID, NFLW  72
     &            MDIM2,MDIM3, MTNODV, MNOD2, MTELS, MTNODP, NDIMN,MNBN, NFLW  73
     &            SOLPN,NDUMP, FDUMP, LNCONS,NF,MDOF,NG,NH,          NFLW  74
     &            NSTEER,MNODLP,NBSID,BVAL,MDOF1,NGNODE,NBFRE,SOL,SOLP, NFLW  75
     &            SOLN,SOLS,INNODE,IXNODE,INLET,BCSDIF,STESNM,       NFLW  76
     &            NPSTIR,NVSTIR,MNODLV,LHALFS,MNQP,MMFAC,MBAND,MQBAND) NFLW  77
C                                                                   NFLW  78
C**** preparation for solving the problem ****                      NFLW  79
```

```
C                                                              NFLW  80
        CALL PREWORK(LINERS,LQNAG,NDIMN,LNCONS,LPLANE,LNEWT,   NFLW  81
     &         LELAST,LQUADC,LQUADS,LBODY,LREZRL,ELDET,MTELS,B,MNODLV,NFLW  82
     &         MNVAR,WORK,MTNODV,SOLN,FLUX,MDIM6,LQUADP,ALMATE,MNODLP,NFLW  83
     &         MDIM,FUNP,MNQP,FUNV,GDV,SOLP,MTNODP,SOLPN,RAD,WGHT,  NFLW  84
     &         NPSTIR,NVSTIR,C,NSTEER,MNSID,LSIMPL,SMATE,SS,VBAR,  NFLW  85
     &         FMATE,CMATE,SOL,DMATE,ELFAC,MMFAC,CCMATA,CCMATB,CCMATC,NFLW  86
     &         CCMATD,CCMATE,CCMATF,AMMATE,LDCOPL,LFRIZT,MTRE1,BODYF, NFLW  87
     &         DD,SYSM,SYSK,MBAND,SYSG,AMMATD,AMATD1,AMATD2,AMATD3,  NFLW  88
     &         AMMAT1,AMMAT2,AMMAT3,NGNODE,MNBN,NH,NBFRE,MDOF1,BVAL, NFLW  89
     &         LTMIND,LFRIZV,LFDBCK,SOLS,INNODE,IXNODE,NG,LPICON,   NFLW  90
     &         COORD,NBSID,BGDP,MDIM3,MDIM2,LMNITR,NELTOP,MNOD2,LLAST,NFLW  91
     &         GEOM,BT,CT,MQBAND,RHSADD,DEL,DELR,RHS,BCSDIF,INLET,  NFLW  92
     &         CCMATG,CCMATH,CCMATI,CCMATJ)                         NFLW  93
C                                                              NFLW  94
C**** transient solution ****                                  NFLW  95
C                                                              NFLW  96
        IF(NSTEP.LT.ITERMX) THEN                                NFLW  97
        CALL TAYGAL(NDIMN,LNCONS,LPLANE,LNEWT,LELAST,LQUADC,   NFLW  98
     &         LQUADS,LBODY,LREZRL,ELDET,MTELS,B,MNODLV,MNVAR,  NFLW  99
     &         WORK,MTNODV,SOLN,FLUX,MDIM6,LQUADP,ALMATE,MNODLP,  NFLW 100
     &         MDIM,FUNP,MNQP,FUNV,GDV,SOLP,MTNODP,SOLPN,RAD,WGHT,  NFLW 101
     &         NPSTIR,NVSTIR,C,NSTEER,MNSID,LSIMPL,SMATE,SS,VBAR,  NFLW 102
     &         FMATE,CMATE,SOL,DMATE,ELFAC,MMFAC,CCMATA,CCMATB,CCMATC,NFLW 103
     &         CCMATD,CCMATE,CCMATF,AMMATE,LDCOPL,LFRIZT,MTRE1,BODYF, NFLW 104
     &         LTMIND,LFRIZV,LFDBCK,SOLS,INNODE,IXNODE,NG,NH,BVAL, NFLW 105
     &         MNBN,MDOF1,NGNODE,NBFRE,SYSK,MBAND,LPICON,LSTGE3,  NFLW 106
     &         COORD,NBSID,BGDP,MDIM3,MDIM2,LLAST,LMNITR,LSTGE2,  NFLW 107
     &         LHALFS,LSECND,RHSADD,DEL,DELR,RHS,SYSM,SYSG,AMMATD,  NFLW 108
     &         AMATD1,AMATD2,AMATD3,AMMAT1,AMMAT2,AMMAT3,BCSDIF,INLET,NFLW 109
     &         CCMATG,CCMATH,CCMATI,CCMATJ)                         NFLW 110
        ENDIF                                                  NFLW 111
C                                                              NFLW 112
C**** output the results ****                                   NFLW 113
C                                                              NFLW 114
        CALL OUTPUT(NDIMN,LSCALE,LMNITR,LLAST,LSTRIM,SOLPN,MTNODP,  NFLW 115
     &         SOLN,MTNODV,MNVAR,NH,NF,MDOF,LELAST,NGNODE,MNBN,NBFRE, NFLW 116
     &         MDOF1,NVSTIR,MNODLV,MTELS,C,B,NPSTIR,MNODLP,NSTEER,  NFLW 117
     &         MNSID,ALMATE,MDIM,SOLP,RAD,WGHT,MNQP,LPLANE,SYSS,  NFLW 118
     &         MBAND,ELFAC,MMFAC,LINERS,LPICON,BVAL,DMATE,FUNP,WORK, NFLW 119
     &         FUNV,GDV,SMATE,SYSQ,MQBAND,NG,SOLQ,SOLS,ELDET,COORD, NFLW 120
     &         NBSID,BGDP,MDIM3,MDIM2,NELTOP,MNOD2)                 NFLW 121
C                                                              NFLW 122
C**** end of program                                           NFLW 123
C                                                              NFLW 124
        STOP                                                   NFLW 125
        END                                                    NFLW 126
```

B.4   Input Instructions

All the input data required by the program are read in the subroutine **INPUTD**. There are two input files, **PARA.DAT** and

GEOM.DAT that a user has to provide. **PARA.DAT** includes all control parameters and **GEOM.DAT** contains the geometry details, the material properties and boundary conditions. In the following, the names of variables and arrays are written in bold faces and a read statement, READ(NIN,*) in the first line of each section is provided for the user to add a comment.

B.4.1 Input instruction for file PARA.DAT

1. Introductory messages

    1.1 READ(NIN,*)

    1.2 READ(NIN,*) **NLINES**

    1.3 READ(NIN,'(80A1)') **ST** (Character string)

    (loop over 1.3 NLINES times)

    where:

    NLINES: Number of lines for comment purpose

2. Logical data

    2.1 READ(NIN,*)

    2.2 READ(NIN,*) **LSCALE, LMNITR, LTMIND, LFRIZT, LFRIZV, LDCOPL, LPLANE**

    2.3 READ(NIN,*)

    2.4 READ(NIN,*) **LNEWT, LSIMPL, LREZRL, LQUADS, LMNITL, LSTGE3, LQUADP**

    2.5 READ(NIN,*)

    2.6 READ(NIN,*) **LNCONS, LSAVR, LSECND, LSTRIM, LQUADC, LBODY, LFDBCK**

    (Ref. Table B.2.1 for descriptions of the above logical parameters)

3. Parameters for flow characteristics

    3.1 READ(NIN,*)

    3.2 READ(NIN,*)**MODEL,NDTYPE**

where:

MODEL : Fluid models, 1 - General power law model; 2 - Elastic model with constant viscosity

NDTYPE : Control parameter, 1 for piecewise constant viscosity case, others for zero Reynolds number

4. Non-Newtonian parameters

   4.1 READ(NIN,*)

   4.2 READ(NIN,*) **PN, POWLAW, EZERO, VISC1, VISC2, EL1, YIELD**

   where:

   PN : Power law index value

   POWLAW : Power law constant μ

   EZERO : cut off value for non-dimensional viscocity

   VISC1 : viscosity μ_1 (cf. chapter 6.4.2)

   VISC2 : viscosity μ_2 (cf. chapter 6.4.2)

   EL1 : relaxation time

   YIELD : yield stress

5. Characteristic parameters/properties

   5.1 READ(NIN,*)

   5.2 READ(NIN,*)**DENSTY, CHRVEL, CHRLEN, VISCTY**   where:

   DENSTY : Mass density

   CHRVEL : Characteristic velocity

   CHRLEN : Characteristic length

   VISCTY : Viscosity

6. Body force vector

   6.1 READ(NIN,*)

   6.2 READ(NIN,*)(**BODYF**(I),I=1,MDIM)

7. Input time-stepping control/physical parameters

    7.1 READ(NIN,*)

    7.2 READ(NIN,*)**ITERMX,TOL,TOLMAS,RELAX**
    where

    ITERMX : Relative time-stepping termination limit

    TOL : Tolerance for time-stepping

    TOLMAS : Tolerance for mass-iteration

    RELAX : Iterative relaxation factor for mass-iteration

8. Time step, time step/averaging factors etc.

    8.1 READ(NIN,*)

    8.2 READ(NIN,*)**DTIM,THETA,NALPHA**
    where:

    DTIM : Time step length

    THETA : Crank-Nicolson factor

    NALPHA : Waiting factor for pressure calculations

9. Domain rotation data

    9.1 READ(NIN,*)

    9.2 READ(NIN,*)ANGLE
    where:

    ANGLE : Domain rotation angle (in degrees)

B.4.2   Input instruction for file GEOM.DAT

1. Introductory messages

    1.1 READ(NIN,*)

    1.2 READ(NIN,*) **NLINES**

    1.3 READ(NIN,9000) **ST** (Character string)
    (loop over 1.3 NLINES times)

where:

NLINES: Number of lines for comment purpose

2. Time step control parameter

    2.1 READ(NIN,*)

    2.2 READ(NIN,*) **NSTEP**

    where:

    NSTEP : 0 for new calculation; ITERMX for calculation of stream function only.

3. Input geometry control data

    3.1 READ(NIN,*)

    3.2 READ(NIN,*) **NTTELS, NTTNOD, NBDNOD, NUMS, NTTDOP**

    where:

    NTTELS : Number of elements

    NTTNOD : Number of nodes

    NBDNOD : Number of boundary nodes

    NUMS : Number of boundary sides specified

    NTTDOP : Number of pressure nodes (vertex nodes)

4. Element connective information

    4.1 READ(NIN,*)

    4.2 READ(NIN,*) **NELNUM, NELTYP, (NELTOP** (J+2,NELNUM), J=1, NVNDEL)

    (loop over 4.2 NTTELS times)

    where:

    NELNUM : Current element number

    NELTYP : Element type, 1= linear triangular elements; 2 = linear Lagrangian elements; 3 = quadratic triangular elements; 4 = quadratic Lagrangian elements; 5 = quadratic Serendipity elements

NELTOP(3 - NVNDEL,NELNUM): Element node connections

5. Nodal Coordinates

5.1 READ(NIN,*)

5.2 READ(NIN,*) **NODNUM,(COORD** (NODNUM, J), J=1, NDIMN)

(loop over 5.2 NTTNOD times)

where:

NODNUM : Current node number

NDIMN : Dimensions of the problem analyzed

COORD($\sim$, J): J: 1 – x, 2 – y for plane case; 1 – r, 2 – z for axial-symmetric case

6. Initial conditions

6.1 READ(NIN,*)

6.2 READ(NIN,*) **NODNUM,(VALUE** (J), J=1, NJDOF1)

(loop over 6.2 NTTNOD times)

where:

NODNUM : Current node number

VALUE(1 $\sim$ 2): Initial nodal velocities

VALUE(3): Pressure

VALUE(4): Initial stream functions

7. Boundary Conditions

7.1 READ(NIN,*)

7.2 READ(NIN,*) **NGNODE** (I),(**NBFRE** (I,J), **BVAL** (I,J),J=1,NDOF1)

(Loop over 7.2 NBDNOD times)

where:

NGNODE(I) : Global node number of current boundary node

NBFRE(I,J): If set 1 then boundary value BVAL(I,J) is assigned, otherwise set 0

BVAL(I,J) : Boundary values: J=1,2 for v_x and v_y in plane problem and for v_r and v_z in axisymmetric case; $J = 3$ for pressure; $J = 4$ for stream function

8*. Optional input data (If **LFDBCK** = .true.)

    8.1* READ(NIN,*)

    8.2* READ(NIN,*) **NBKNOD**

    8.3* READ(NIN,*)

    8.4* READ(NIN,*) (**INNODE** (I),I=1,NBKNOD)

    8.5* READ(NIN,*)

    8.6* READ(NIN,*) (**IXNODE** (I),I=1,NBKNOD)

    where

    NBKNOD : Number of inlet nodes for feed back

    INNODE(I) : Node Number on inlet for feed back

    IXNODE(I) : Node Number on outlet for feed back

9* Optional input data (IF LTMIND=.FALSE.)

    9.1* READ(NIN,*)

    9.2* READ(NIN,*) **MAXM**

    9.3* READ(NIN,*)

    9.4* READ(NIN,*) **NINLET**

    9.5* READ(NIN,*)

    9.6* READ(NIN,*) (**INLET** (I),I=1,NINLET)

    9.7* READ(NIN,*)

    9.8* READ(NIN,*)((**BCSDIF** (I,J),I=1,NINLET),J=1,NVET)

    where

    MAXM : Number of time steps for input of time dependent boundary condtions

    NINLET: Number of nodes on inlet

INLET(I) : Node Number on inlet

BCSDIF(I,J): Differences on inlet

The input subroutine, INPUTD, is listed in the following lines.

```
      SUBROUTINE INPUTD (LPICON,LLAST,LNEWT,LQUADS,LQNAG,LQUADP,LSECND, INPU  1
     &           LSCALE,LTMIND,LMNITR,LSTRIM,LSIMPL,LQUADC,LPLANE,        INPU  2
     &           LELAST,LBODY,LREZRL,LINERS,LFDBCK,LSAVR,LSTGE2,          INPU  3
     &           LSTGE3,LDCOPL,LFRIZV,LFRIZT,MNVAR,BODYF,MDIM,MONS,MONP,  INPU  4
     &           MONV,NSCBRY,MMONV, MMONP, CTSTEP,  BGDP,DUMPNM, STARNM,  INPU  5
     &           VECTNM, PRENM, PATNM,COORD, NELTOP, MCTSTP, MNSID,       INPU  6
     &           MDIM2,MDIM3, MTNODV, MNOD2, MTELS, MTNODP, NDIMN,MNBN,   INPU  7
     &           SOLPN,NDUMP, FDUMP, LNCONS,NF,MDOF,NG,NH,                INPU  8
     &           NSTEER,MNODLP,NBSID,BVAL,MDOF1,NGNODE,NBFRE,SOL,SOLP,    INPU  9
     &           SOLN,SOLS,INNODE,IXNODE,INLET,BCSDIF,STESNM,             INPU 10
     &           NPSTIR,NVSTIR,MNODLV,LHALFS,MNQP,MMFAC,MBAND,MQBAND)     INPU 11
C****************************************************************************INPU 12
C                                                                          INPU 13
C     Input all data                                                       INPU 14
C                                                                          INPU 15
C****************************************************************************INPU 16
      IMPLICIT DOUBLE PRECISION (A-H,O-Z)                                  INPU 17
      INCLUDE 'NFLOW.FIN'                                                  INPU 18
C     ====================                                                 INPU 19
      CHARACTER*80TITLE,ST                                                 INPU 20
      CHARACTER*20DUMPNM,STESNM,VECTNM,PRENM,PATNM,STARNM                  INPU 21
      LOGICAL LSCALE,LMNITR,LTMIND,LELAST,LFRIZT,LFRIZV,                   INPU 22
     &        LDCOPL,LPLANE,LNEWT,LSIMPL,LREZRL,LQUADS,LINERS,             INPU 23
     &        LQNAG,LSTGE3,LQUADP,LNCONS,LSAVR,LSECND,LSTRIM,              INPU 24
     &        LSTGE2,LQUADC,LBODY,LFDBCK,LPICON,LLAST,LHALFS               INPU 25
      DIMENSION TCD(2),BODYF(MDIM),CTSTEP(0:MCTSTP),                       INPU 26
     .          MONV(MMONV),MONP(MMONP),MONS(MMONV),                       INPU 27
     .          NSCBRY(MMONV),NELTOP(MNOD2,MTELS),SOLPN(MTNODP),           INPU 28
     .          COORD(MTNODV,MDIM),NBSID(MNSID,MDIM3),                     INPU 29
     .          BGDP(MNSID,MDIM2), SOLN(MTNODV,MNVAR)                      INPU 30
      DIMENSION NG(MTNODV),NH(MTNODV)                                      INPU 31
      DIMENSION BVAL(MNBN,MDOF1),NGNODE(MNBN),NBFRE(MNBN,MDOF1),           INPU 32
     &          SOL(MTNODV,MNVAR),SOLP(MTNODP),                           INPU 33
     &          SOLS(MTNODP),NSTEER(MNODLP,MNSID),                        INPU 34
     &          NPSTIR(MNODLP,MTELS),NVSTIR(MNODLV,MTELS)                  INPU 35
      DIMENSION NF(MTNODV,MDOF)                                           INPU 36
      DIMENSION INNODE(*),IXNODE(*),INLET(*),BCSDIF(MNBN,MNVAR)            INPU 37
C                                                                          INPU 38
C*** Input - output channel numbers                                        INPU 39
C                                                                          INPU 40
      NIN=5                                                                INPU 41
      NOUT=6                                                               INPU 42
C                                                                          INPU 43
C*** Set itest for full checking                                           INPU 44
C                                                                          INPU 45
      ITEST=0                                                              INPU 46
C                                                                          INPU 47
C**** open file channels ****                                              INPU 48
C                                                                          INPU 49
      OPEN(UNIT=NIN,FILE='PARA.DAT')                                       INPU 50
```

```
      OPEN(UNIT=NOUT,FILE='RES.DAT')                                 INPU  51
C                                                                    INPU  52
C*  INITIALISE PARAMETERS                                            INPU  53
C                                                                    INPU  54
      CALL PRMTRST(MDIM,MNQP,MDOF,MNODLP,MNODLV,MMFAC,MNVAR,          INPU  55
     &             MTNODV,MNOD2,MTNODP,MTELS,MBAND,MQBAND,            INPU  56
     &             LHALFS,LPICON,LLAST,LELAST,LINERS,LQNAG,LSTGE2)    INPU  57
C                                                                    INPU  58
      IWORK3=IWORK1                                                   INPU  59
C                                                                    INPU  60
C*** First output an introductory message                           INPU  61
C                                                                    INPU  62
      READ(NIN,*)                                                    INPU  63
      READ(NIN,*)NLINES                                              INPU  64
      DO 10 I = 1, NLINES                                            INPU  65
      READ(NIN,9000)ST                                               INPU  66
      IF (ST(1:1).EQ.'#') THEN                                       INPU  67
      DO  5 J=1,80                                                   INPU  68
      TITLE(J:J)=ST(J:J)                                             INPU  69
  5   CONTINUE                                                       INPU  70
      ENDIF                                                          INPU  71
      WRITE (NOUT, 9000) ST                                          INPU  72
 10   CONTINUE                                                       INPU  73
      WRITE(NOUT,9000)                                               INPU  74
C                                                                    INPU  75
C*** Input - logical flags/parameters                               INPU  76
C                                                                    INPU  77
      READ(NIN,*)                                                    INPU  78
      READ(NIN,*)LSCALE,LMNITR,LTMIND,LFRIZT,LFRIZV,LDCOPL,LPLANE    INPU  79
      READ(NIN,*)                                                    INPU  80
      READ(NIN,*)LNEWT,LSIMPL,LREZRL,LQUADS,LSTGE3,LQUADP,LELAST     INPU  81
      READ(NIN,*)                                                    INPU  82
      READ(NIN,*)LNCONS,LSAVR,LSECND,LSTRIM,LQUADC,LBODY,LFDBCK      INPU  83
C                                                                    INPU  84
C* OPEN CHANNEL FOR STREAM FUNCTION.                                 INPU  85
C                                                                    INPU  86
      IF(LSTRIM) THEN                                                INPU  87
      OPEN(UNIT=55,FILE='TSTPLT.INP')                                INPU  88
      ENDIF                                                          INPU  89
C                                                                    INPU  90
C*** Output - logical flags/paramaters                              INPU  91
C                                                                    INPU  92

      CALL      LOGOUT(LNEWT, LQUADS,LNCONS,LPICON,LQNAG,            INPU  93
     *                LSECND,LSCALE,LTMIND,LMNITR,LSTRIM,            INPU  94
     *                LSIMPL,LQUADC,LPLANE,LELAST,LLAST, LBODY,      INPU  95
     *                LINERS,LFDBCK,LSAVR,LSTGE3,LDCOPL,LFRIZV,LFRIZT,INPU 96
     *                LQUADP,LREZRL,LSTGE2)                          INPU  97
C                                                                    INPU  98
C*** parameters for flow characteristics                            INPU  99
C                                                                    INPU 100
      READ(NIN,*)                                                    INPU 101
      READ(NIN,*)MODEL,NDTYPE                                        INPU 102
      WRITE (NOUT,9010)MODEL,NDTYPE                                  INPU 103
C                                                                    INPU 104
C*** non-Newtonian parameters                                       INPU 105
C                                                                    INPU 106
```

```
      READ(NIN,*)                                               INPU 107
      READ(NIN,*)PN,POWLAW,EZERO,VISC1,VISC2,EL1,YIELD           INPU 108
      WRITE (NOUT,9015)PN,POWLAW,EZERO,VISC1,VISC2,EL1,YIELD     INPU 109
      EINF=0.0DO                                                 INPU 110
      RLAM=0.0DO                                                 INPU 111
C                                                                INPU 112
C*** characteristic parameters/properties                       INPU 113
C                                                                INPU 114
      READ(NIN,*)                                                INPU 115
      READ(NIN,*)DENSTY,CHRVEL,CHRLEN,VISCTY                     INPU 116
      WRITE (NOUT,9020)DENSTY,CHRVEL,CHRLEN,VISCTY               INPU 117
      CHRVELI=1.DO/CHRVEL                                        INPU 118
      CHRLV=CHRLEN*CHRVEL                                        INPU 119
      CHRLVI=1.DO/CHRLV                                          INPU 120
      DENCV2=DENSTY*CHRVEL**2                                    INPU 121
      DENCV2I=1.DO/DENCV2                                        INPU 122
C                                                                INPU 123
C*** shear rate limit for the power law model                   INPU 124
C                                                                INPU 125
C     SCINV0=(VISCTY/POWLAW)**(1.DO/(PN-1.DO))                   INPU 126
C                                                                INPU 127
C*** body force vector                                          INPU 128
C                                                                INPU 129
      READ(NIN,*)                                                INPU 130
      READ(NIN,*)(BODYF(I),I=1,MDIM)                             INPU 131
      WRITE(NOUT,9022) (BODYF(I),I=1,MDIM)                       INPU 132
C                                                                INPU 133
C*** Define non-dimensional parameters                          INPU 134
C                                                                INPU 135
      IF(LELAST)VISCTY=VISC1+VISC2                               INPU 136
      REYNLD=CHRVEL*CHRLEN*DENSTY/VISCTY                         INPU 137
      IF(.NOT.LREZRL)NDTYPE=2                                    INPU 138
      IF(NDTYPE.EQ.1)THEN                                        INPU 139
      REYNLD1=REYNLD                                             INPU 140
      REYNLD2=1.DO                                               INPU 141
C                                                                INPU 142
C*** piecewise constant viscosity case                          INPU 143
C                                                                INPU 144
      IF(LSAVR)REYNLD1=CHRVEL*CHRLEN*DENSTY                      INPU 145
      ELSE                                                       INPU 146
C                                                                INPU 147
C*** zero Reynolds No. considered here!                         INPU 148
C                                                                INPU 149
      REYNLD1=1.DO                                               INPU 150
      REYNLD2=REYNLD                                             INPU 151
C                                                                INPU 152
C*** piecewise constant viscosity case                          INPU 153
C                                                                INPU 154
      IF(LSAVR)REYNLD2=CHRVEL*CHRLEN*DENSTY                      INPU 155
      ENDIF                                                      INPU 156
      CONST1=-VISC1/EL1                                          INPU 157
      WRITE (NOUT,9030) REYNLD1,REYNLD2,CONST1                   INPU 158
C                                                                INPU 159
C*** Input time-stepping control/physical parameters            INPU 160
C                                                                INPU 161
      MTER=1                                                     INPU 162
      NTER=1                                                     INPU 163
```

```
      READ(NIN,*)                                                INPU 164
      READ(NIN,*)ITERMX,TOL,TOLMAS,RELAX                         INPU 165
C      READ(NIN,*)ITERMX,NTER,MTER,TOL,TOLMAS,RELAX              INPU 166
C      WRITE(NOUT,9040) ITERMX,NTER,MTER,TOL,TOLMAS,RELAX        INPU 167
      WRITE(NOUT,9040) ITERMX,TOL,TOLMAS,RELAX                   INPU 168
C                                                                INPU 169
C*** Time step, time step/averaging factors etc.                INPU 170
C                                                                INPU 171
      READ(NIN,*)                                                INPU 172
      READ(NIN,*)DTIM,THETA,NALPHA                               INPU 173
      IF(NALPHA.EQ.0)THEN                                        INPU 174
      THETA=1.D0                                                 INPU 175
      ENDIF                                                      INPU 176
      CONST2=2.D0*VISC1/DTIM                                     INPU 177
      WRITE (NOUT,9050) DTIM,THETA,NALPHA,CONST2                 INPU 178
C                                                                INPU 179
C*** Basic geometric parameters                                 INPU 180
C                                                                INPU 181
      READ(NIN,*)                                                INPU 182
      READ(NIN,*)ANGLE                                           INPU 183
      WRITE (NOUT, 9051) ANGLE                                   INPU 184
      ANGLE=ANGLE*3.141592654D0/180.D0                           INPU 185
C                                                                INPU 186
C*** Close PARA.DAT and open GEOM.DAT                            INPU 187
C                                                                INPU 188
      CLOSE(NIN)                                                 INPU 189
      OPEN (NIN, FILE='GEOM.DAT')                                INPU 190
C                                                                INPU 191
C*** First read an introductory message on GEOM.DAT             INPU 192
C                                                                INPU 193
      READ(NIN,*)                                                INPU 194
      READ(NIN,*)NLINES                                          INPU 195
      DO 30 I = 1, NLINES                                        INPU 196
      READ(NIN,9000)ST                                           INPU 197
30    CONTINUE                                                   INPU 198
      READ(NIN,*)                                                INPU 199
      READ(NIN,*)NSTEP                                           INPU 200
C                                                                INPU 201
C*** Input of nodal geometry                                    INPU 202
C                                                                INPU 203


      READ(NIN,*)                                                INPU 204
      READ(NIN,*)NTTELS,NTTNOD,NBDNOD,NUMS,NTTDOP                INPU 205
      WRITE (NOUT,9060) NTTELS,NTTNOD,NBDNOD,NUMS,NTTDOP         INPU 206
C                                                                INPU 207
C*** Error checking on problem dependent global parameters      INPU 208
C                                                                INPU 209
      IF(NTTELS.GT.MTELS)THEN                                    INPU 210
      WRITE (NOUT,9061) NTTELS,MTELS                             INPU 211
      STOP                                                       INPU 212
      ENDIF                                                      INPU 213
      IF(NTTNOD.GT.MTNODV)THEN                                   INPU 214
      WRITE (NOUT,9062) NTTNOD,MTNODV                            INPU 215
      STOP                                                       INPU 216
      ENDIF                                                      INPU 217
      IF(NBDNOD.GT.MNBN)THEN                                     INPU 218
      WRITE (NOUT,9063) NBDNOD,MNBN                              INPU 219
```

```
          STOP                                                      INPU 220
          ENDIF                                                     INPU 221
          IF(NUMS.GT.MNSID)THEN                                     INPU 222
          WRITE (NOUT,9064) NUMS,MNSID                              INPU 223
          STOP                                                      INPU 224
          ENDIF                                                     INPU 225
          IF(NTTDOP.GT.MTNODP)THEN                                  INPU 226
          WRITE (NOUT,9065) NTTDOP,MTNODP                           INPU 227
          STOP                                                      INPU 228
          ENDIF                                                     INPU 229
C                                                                   INPU 230
C*** Element Topology                                              INPU 231
C                                                                   INPU 232
          READ(NIN,*)                                               INPU 233
          DO 1020 I=1,NTTELS                                        INPU 234
          READ(NIN,*)NELNUM,NELTYP,(NELTOP(J+2,NELNUM),J=1,NVNDEL)  INPU 235
          NELTOP(1,NELNUM)=NELTYP                                   INPU 236
          NELTOP(2,NELNUM)=NVNDEL                                   INPU 237
 1020     CONTINUE                                                  INPU 238
          IF(.NOT.LMNITR)THEN                                       INPU 239
          WRITE(NOUT,9070)                                          INPU 240
          DO 1021 I=1,NTTELS                                        INPU 241
          WRITE(NOUT,9080)I,NELTYP,NVNDEL,                          INPU 242
      *                   (NELTOP(J+2,I),J=1,NVNDEL)                INPU 243
 1021     CONTINUE                                                  INPU 244
          ENDIF                                                     INPU 245
C                                                                   INPU 246
C*** Nodal Coordinates                                             INPU 247
C                                                                   INPU 248
          READ(NIN,*)                                               INPU 249
          IF(.NOT.LMNITR)WRITE(NOUT,9090)                           INPU 250
          DO 1040 I=1,NTTNOD                                        INPU 251
          READ(NIN,*)NODNUM,(COORD(NODNUM,J),J=1,NDIMN)             INPU 252
          IF(LPLANE)THEN                                            INPU 253
          IF(DABS(ANGLE).LT.DSMALL)GOTO 1026                        INPU 254
          ENDIF                                                     INPU 255
C                                                                   INPU 256
C***  2D coordinate translation                                    INPU 257
C                                                                   INPU 258
          DO 1025 J=1,NDIMN                                         INPU 259
          TCD(J)=COORD(NODNUM,J)                                    INPU 260
 1025     CONTINUE                                                  INPU 261
C                                                                   INPU 262
C*** for axisymmetry X -> R, Y -> Z in usual convention            INPU 263
C                                                                   INPU 264
          COORD(NODNUM,1)=TCD(1)*COS(ANGLE)-TCD(2)*SIN(ANGLE)       INPU 265
          COORD(NODNUM,2)=TCD(1)*SIN(ANGLE)+TCD(2)*COS(ANGLE)       INPU 266
 1026     CONTINUE                                                  INPU 267
          IF(.NOT.LMNITR)THEN                                       INPU 268
          WRITE(NOUT,9100)NODNUM,(COORD(NODNUM,J),J=1,NDIMN)        INPU 269
          ENDIF                                                     INPU 270
C                                                                   INPU 271
C*** Non-dimensionalize if appropriate                             INPU 272
C                                                                   INPU 273
          IF(.NOT.LSCALE)THEN                                       INPU 274
          DO 1030 J=1,NDIMN                                         INPU 275
          COORD(NODNUM,J)=COORD(NODNUM,J)/CHRLEN                     INPU 276
```

```
1030  CONTINUE                                                  INPU 277
      ENDIF                                                     INPU 278
1040  CONTINUE                                                  INPU 279
C                                                               INPU 280
C* WRITE OUT NODAL AND ELEMENT INFORMATION FOR STREAM FUNCTION  INPU 281
C                                                               INPU 282
      IF(LSTRIM)CALL OUTSTRM(1,55,NTTNOD,NTTELS,NVNDEL,3,NDIMN,COORD,  INPU 283
     & MTNODV,MDIM,NELTOP,MNOD2,MTELS,SOLS,MTNODP,ISOLP,SOLN,MNVAR)    INPU 284
C                                                               INPU 285
C*** Initial and boundary conditions                           INPU 286
C                                                               INPU 287
      CALL NFSET(NELTOP,MNOD2,MTELS,NF,MTNODV,MDOF)             INPU 288
      CALL CONFRE(NF,MTNODV,MDOF,NG,NH,LMNITR,                  INPU 289
     &           NELTOP,MNOD2,MTELS,NPSTIR,MNODLP,NVSTIR,MNODLV) INPU 290
C                                                               INPU 291
C   nullify solution vectors:  velocities and pressure         INPU 292
C                                                               INPU 293
      CALL NULMAT(SOLN,ISOL,JSOL,NTTDOV,JSOL,ITEST)            INPU 294
      CALL NULVEC(SOLPN,ISOLP,NTTDOP,ITEST)                    INPU 295
C                                                               INPU 296
C*** read initial conditions and transcribe to SOLN,SOLPN & SOL,SOLP  INPU 297
C                                                               INPU 298


      READ(NIN,*)                                               INPU 299
      IF(.NOT.LMNITR)WRITE(NOUT,9110)                          INPU 300
      CALL INPSCA(CHRVELI,DENCV2I,CHRLVI,LSCALE,LMNITR,NDIMN,  INPU 301
     &           NG,MTNODV,NH,SOLN,MNVAR,SOLPN,MTNODP,NF,MDOF,SOL,SOLP) INPU 302
C                                                               INPU 303
C*** Read Boundary Conditions                                  INPU 304
C                                                               INPU 305
      READ(NIN,*)                                               INPU 306
      IF(.NOT.LMNITR)WRITE(NOUT,9120)                          INPU 307
C                                                               INPU 308
C*** read in boundary information and  transcribe to SOL*      INPU 309
C                                                               INPU 310
      IF(NBDNOD.GT.0)THEN                                       INPU 311
      CALL INPBRY(NDIMN,BVAL,MNBN,MDOF1,NGNODE,NBFRE,SOL,MTNODV,  INPU 312
     &           MNVAR,SOLP,MTNODP,SOLPN,SOLN,SOLS,LSCALE,LMNITR,NG,NH)  INPU 313
      ELSE                                                      INPU 314
      WRITE (NOUT, *) 'NBDNOD <= 0'                            INPU 315
      ENDIF                                                     INPU 316
C                                                               INPU 317
C*** Read Boundary Side Information                            INPU 318
C                                                               INPU 319
C     READ(NIN,*)                                               INPU 320
C     DO 1050 I=1,NUMS                                          INPU 321
C     READ(NIN,*)(NBSID(I,J),J=1,MDIM3),                       INPU 322
C   *                (BGDP(I,K),K=1,MDIM2)                     INPU 323
      DO I=1,NUMS                                               INPU 324
      DO J=1,MDIM3                                              INPU 325
      NBSID(I,J)=0.0                                            INPU 326
      ENDDO                                                     INPU 327
      DO J=1,MDIM2                                              INPU 328
      BGDP(I,J)=0.0                                             INPU 329
      ENDDO                                                     INPU 330
      ENDDO                                                     INPU 331
C                                                               INPU 332
```

```
C1050  CONTINUE                                                        INPU 334
       IF(.NOT.LMNITR)THEN                                             INPU 335
       WRITE(NOUT,9130)                                                INPU 336
       DO 1051 I=1,NUMS                                                INPU 337
       WRITE (NOUT,9140)I,(NBSID(I,J),J=1,MDIM3),                      INPU 338
     *                  (BGDP(I,K),K=1,MDIM2)                          INPU 339
1051   CONTINUE                                                        INPU 340
       ENDIF                                                           INPU 341
C                                                                      INPU 342
C*** now construct steering vector for all boundary sides             INPU 343
C                                                                      INPU 344
       CALL GETDIS(NSTEER,MNODLP,NBSID,MNSID,MDIM3)                    INPU 345
C                                                                      INPU 346
C*** Feed Back/Time-dependent b.c Information                         INPU 347
C                                                                      INPU 348
       IF(LFDBCK)                                                      INPU 349
     & CALL INPXTR(LFDBCK,LMNITR,LTMIND,INNODE,IXNODE,INLET,BCSDIF,    INPU 350
     &             MNBN,MNVAR)                                         INPU 351
       IF(.NOT.LTMIND)                                                 INPU 352
     & CALL INPXTR(LFDBCK,LMNITR,LTMIND,INNODE,IXNODE,INLET,BCSDIF,    INPU 353
     &             MNBN,MNVAR)                                         INPU 354
       RETURN                                                          INPU 355
9000   FORMAT (A80)                                                    INPU 356
9010   FORMAT('Model specified  : ',I3,' Non-dimen type :',I3/)       INPU 357
9015   FORMAT('Powerlaw index   : ',D12.5, ' ',                       INPU 358
     *        /'Powerlaw const.  : ',D12.5, ' ',                       INPU 359
C    *        /'Carreau EINF     : ',D12.5, ' ',                       INPU 360
     *        /'Carreau EZERO    : ',D12.5, ' ',                       INPU 361
C    *        /'Carreau RLAM     : ',D12.5, ' ',                       INPU 362
     *        /'viscosity1 VISC1 : ',D12.5, ' ',                       INPU 363
     *        /'viscosity2 VISC2 : ',D12.5, ' ',                       INPU 364
     *        /'relax. time EL1  : ',D12.5, ' ',                       INPU 365
     *        /'yield stress     : ',D12.5, ' ',/)                     INPU 366
9020   FORMAT('Density          : ',D12.5, ' ',                       INPU 367
     *        /'Char.velocity    : ',D12.5, ' ',                       INPU 368
     *        /'Char.length      : ',D12.5, ' ',                       INPU 369
     *        /'Viscosity        : ',D12.5, ' ',/)                     INPU 370
9022   FORMAT('Body force vector: ',D12.5,2(2X,D12.5), ' ',/)         INPU 371
9030   FORMAT('Reynolds No. 1   : ',D12.5, ' ',                       INPU 372
     *        /'Reynolds No. 2   : ',D12.5, ' ',                       INPU 373
     *        /'extra parm CONST1: ',D12.5, ' ' )                      INPU 374
9040   FORMAT('Relative time-stepping termination limit : ',I7,       INPU 375
C    *        /'Mass-itn termination limit               : ',I7,       INPU 376
C    *        /'Mass-itn termination limit (vels only!!) : ',I7,       INPU 377
     *        /'Tolerance for time-stepping              : ',D12.5,    INPU 378
     *        /'Tolerance for Mass-itn                   : ',D12.5,    INPU 379
     *        /'Iterative relaxation factor for Mass-itn : ',D12.5,/)  INPU 380
9050   FORMAT('Time-step                : ',D12.5,                     INPU 381
     *        /'Crank-N factor           : ',D12.5,                     INPU 382
     *        /'Alpha                    : ',I3,                        INPU 383
     *        /'extra parm CONST2        : ',D12.5)                     INPU 384
9051   FORMAT ('Domain rotation : ',D12.5, '   degrees')              INPU 385
9060   FORMAT('Number of elements       : ',i5,                       INPU 386
     *        /'Number of nodes          : ',i5,                        INPU 387
     *        /'Number of boundary nodes : ',i5,                        INPU 388
     *        /'Number of boundary sides : ',i5,                        INPU 389
     *        /'Number of vertex nodes   : ',i5,/)                      INPU 390
```

```
9061  FORMAT('INPUTD: Error in number of elements specified: ',        INPU 391
   *         'NTTELS = ',i5,'MTELS = ',i5)                              INPU 392
9062  FORMAT('INPUTD: Error in number of nodes specified: ',           INPU 393
   *         'NTTNOD = ',i5,'MTNODV = ',i5)                            INPU 394
9063  FORMAT('INPUTD: Error in number of boundary nodes specified: ',  INPU 395
   *         'NBDNOD = ',i5,'MNBN = ',i5)                              INPU 396
9064  FORMAT('INPUTD: Error in number of boundary sides specified: ',  INPU 397
   *         'NUMS = ',i5,'MNSID = ',i5)                               INPU 398
9065  FORMAT('INPUTD: Error in number of pressure nodes specified: ',  INPU 399
   *         'NTTDOP = ',i5,'MTNODP = ',i5)                            INPU 400
9070  FORMAT(//'Element Topology :'//                                  INPU 401
   *               '  Elmt  Type  Nodes.....')                         INPU 402
9080  FORMAT(13I6)                                                     INPU 403
9090  FORMAT(///'Nodal Coordinates :'//'  Node  Coordinates......')    INPU 404
9100  FORMAT(I6,3(D12.5, ' '))                                         INPU 405
9110  FORMAT(///'Initial conditions:-'/)                               INPU 406
9120  FORMAT(///'Boundary conditions:-'/)                              INPU 407
9130  FORMAT(///'Boundary Side Information:-'//                         INPU 408
   *            '  Side    Nodes',5X,'NBSID.....',8X,'BGDP.....')      INPU 409
9140  FORMAT(I5,5I5,3X,6F10.5)                                         INPU 410
      END                                                              INPU 411
```

B.5   Documented Example

The example for the flow over a step is considered here again with
the geometry, boundary and loading conditions shown in Figure
4.2a. The second adaptive mesh, shown in Figure 6.3b, is used in
this example for the reason of less nodal and element numbers produced. The power law index equals to 0.75 and the characteristic
length and velocity used are unit. The normalized parameter $\bar{\mu}$ is
0.02, corresponding to the nominal Reynolds number of 50. The
calculation is convergenced at time step 4055 for the convergency
tolerance 10^{-8} and time increment of 0.1. The input files and results are listed at below. To keep the print out of theses files within
the page boundary, some format has been altered and spaces have
been shortened.

Input Data File PARA.DAT

Contents of PARA.DAT which contains mainly controlling parameters, are listed as follow:

```
NUMBER OF TEXT LINES TO PRINT
7
```

```
*******************************************************

# Flow over a step for D=1, L=4D STEP =0.4DX0.4D
# CONSTANT VELOCITY AT INLET (unit).
# (phi=0 at Y=0, phi=1 at y=1  )
# for adaptive mesh B
*******************************************************
LSCALE, LMNITR, LTMIND, LFRIZT, LFRIZV, LDCOPL, LPLANE
t       T       T       F       F       F       t
LNEWT,  LSIMPL, LREZRL, LQUADS, LSTGE3, LQUADP, LELAST
f       T       T       F       T       F       T
LNCONS, LSAVR,  LSECND, LSTRIM, LQUADC, LBODY,  LFDBCK
T       F       T       T       F       F       F
MODEL NDTYPE
1       1
PN POWLAW EZERO   VISC1    VISC2    EL1    YIELD
0.75 0.02    0.2     0.8888   0.1112    1.0    1.
DENSTY  CHRVEL  CHRLEN   VISCTY
1.0     1.0     1.0      1.0
BODYF1  BODYF2  BODYF3
0.0     0.0     0.0
ITERMX  TOL       TOLMAS     RELAX
10000     1.E-8    0.01        0.7
DTIM  THETA  ALPHA
0.1 0.5 1
ANGLE
0.0
```

Input Data File GEOM.DAT

Content of GEOM.DAT which contains geometry, boundary and loading conditions are listed as follow:

```
NUMBER OF TEXT LINES TO PRINT
    5
*************************************
```

```
#Flow over a step

****************************************
NSTEP
   0
TOTELS   TOTNOD   BNDNOD   NUMS   TOTDOP
 186       427      108      0     121
ELEMENTS :
1     3     4     17    55    123    124    122
  2   3    16      4    56    126    127    125
  3   3    15     16    56    127    129    128
  4   3    17     18    55    131    123    130
  5   3     3     15    57    133    134    132
  6   3    18      5    19    136    137    135
      . . .
      . . .
182   3    27     28   121    422    423    421
183   3    28      6    29    425    418    424
184   3    27    121   113    426    403    423
185   3   113    121   120    427    420    426
186   3   120    121    28    422    417    427
COORDINATES :
    1    0.0000    0.0000
    2    1.2000    0.0000
    3    1.2000    0.4000
    4    1.6000    0.4000
    5    1.6000    0.0000
    6    4.0000    0.0000
      . . .
      . . .
  421    3.5645    0.0000
  422    3.6342    0.0760
  423    3.4903    0.0760
  424    3.8542    0.0000
  425    4.0000    0.1484
  426    3.4776    0.2246
  427    3.6377    0.2315
INITIAL CONDITIONS :
```

```
  1   0.0000E+00   0.0000E+00   0.0000E+00   0.0000E+00
  2   0.0000E+00   0.0000E+00   0.0000E+00   0.0000E+00
  3   0.0000E+00   0.0000E+00   0.0000E+00   0.0000E+00
  4   0.0000E+00   0.0000E+00   0.0000E+00   0.0000E+00
  5   0.0000E+00   0.0000E+00   0.0000E+00   0.0000E+00
  6  .0.0000E+00   0.0000E+00   0.0000E+00   0.0000E+00

    . . .
    . . .

423   0.0000E+00   0.0000E+00   0.0000E+00   0.0000E+00
424   0.0000E+00   0.0000E+00   0.0000E+00   0.0000E+00
425   0.0000E+00   0.0000E+00   0.0000E+00   0.0000E+00
426   0.0000E+00   0.0000E+00   0.0000E+00   0.0000E+00
427   0.0000E+00   0.0000E+00   0.0000E+00   0.0000E+00
BOUNDARY CONDITIONS :
  1 1   0.0000 1   0.0000 0   0.0000 1   0.0000
  9 1   0.0000 1   0.0000 0   0.0000 1   0.0000
 10 1   0.0000 1   0.0000 0   0.0000 1   0.0000
 11 1   0.0000 1   0.0000 0   0.0000 1   0.0000
 12 1   0.0000 1   0.0000 0   0.0000 1   0.0000
 13 1   0.0000 1   0.0000 0   0.0000 1   0.0000
  2 1   0.0000 1   0.0000 0   0.0000 1   0.0000
278 1   0.0000 1   0.0000 0   0.0000 1   0.0000
281 1   0.0000 1   0.0000 0   0.0000 1   0.0000
290 1   0.0000 1   0.0000 0   0.0000 1   0.0000
315 1   0.0000 1   0.0000 0   0.0000 1   0.0000
352 1   0.0000 1   0.0000 0   0.0000 1   0.0000
362 1   0.0000 1   0.0000 0   0.0000 1   0.0000
 14 1   0.0000 1   0.0000 0   0.0000 1   0.0000
  3 1   0.0000 1   0.0000 0   0.0000 1   0.0000
282 1   0.0000 1   0.0000 0   0.0000 1   0.0000
311 1   0.0000 1   0.0000 0   0.0000 1   0.0000
 15 1   0.0000 1   0.0000 0   0.0000 1   0.0000
 16 1   0.0000 1   0.0000 0   0.0000 1   0.0000
  4 1   0.0000 1   0.0000 0   0.0000 1   0.0000
125 1   0.0000 1   0.0000 0   0.0000 1   0.0000
128 1   0.0000 1   0.0000 0   0.0000 1   0.0000
132 1   0.0000 1   0.0000 0   0.0000 1   0.0000
 17 1   0.0000 1   0.0000 0   0.0000 1   0.0000
```

```
 18 1   0.0000 1   0.0000 0   0.0000 1   0.0000
  5 1   0.0000 1   0.0000 0   0.0000 1   0.0000
122 1   0.0000 1   0.0000 0   0.0000 1   0.0000
130 1   0.0000 1   0.0000 0   0.0000 1   0.0000
135 1   0.0000 1   0.0000 0   0.0000 1   0.0000
 19 1   0.0000 1   0.0000 0   0.0000 1   0.0000
 20 1   0.0000 1   0.0000 0   0.0000 1   0.0000
 21 1   0.0000 1   0.0000 0   0.0000 1   0.0000
 22 1   0.0000 1   0.0000 0   0.0000 1   0.0000
 23 1   0.0000 1   0.0000 0   0.0000 1   0.0000
 24 1   0.0000 1   0.0000 0   0.0000 1   0.0000
 25 1   0.0000 1   0.0000 0   0.0000 1   0.0000
 26 1   0.0000 1   0.0000 0   0.0000 1   0.0000
 27 1   0.0000 1   0.0000 0   0.0000 1   0.0000
 28 1   0.0000 1   0.0000 0   0.0000 1   0.0000
  6 1   0.0000 1   0.0000 1   0.0000 1   0.0000
136 1   0.0000 1   0.0000 0   0.0000 1   0.0000
264 1   0.0000 1   0.0000 0   0.0000 1   0.0000
269 1   0.0000 1   0.0000 0   0.0000 1   0.0000
272 1   0.0000 1   0.0000 0   0.0000 1   0.0000
284 1   0.0000 1   0.0000 0   0.0000 1   0.0000
305 1   0.0000 1   0.0000 0   0.0000 1   0.0000
336 1   0.0000 1   0.0000 0   0.0000 1   0.0000
369 1   0.0000 1   0.0000 0   0.0000 1   0.0000
402 1   0.0000 1   0.0000 0   0.0000 1   0.0000
421 1   0.0000 1   0.0000 0   0.0000 1   0.0000
424 1   0.0000 1   0.0000 0   0.0000 1   0.0000
 29 0   0.0000 1   0.0000 0   0.0000 0   0.0000
 30 0   0.0000 1   0.0000 0   0.0000 0   0.0000
 31 0   0.0000 1   0.0000 0   0.0000 0   0.0000
  7 1   0.0000 1   0.0000 0   0.0000 1   1.0000
328 0   0.0000 1   0.0000 0   0.0000 1   1.0000
379 0   0.0000 1   0.0000 0   0.0000 1   1.0000
407 0   0.0000 1   0.0000 0   0.0000 1   1.0000
425 0   0.0000 1   0.0000 0   0.0000 1   1.0000
 32 1   0.0000 1   0.0000 0   0.0000 1   1.0000
 33 1   0.0000 1   0.0000 0   0.0000 1   1.0000
 34 1   0.0000 1   0.0000 0   0.0000 1   1.0000
```

```
 35 1   0.0000 1   0.0000 0   0.0000 1   1.0000
 36 1   0.0000 1   0.0000 0   0.0000 1   1.0000
 37 1   0.0000 1   0.0000 0   0.0000 1   1.0000
 38 1   0.0000 1   0.0000 0   0.0000 1   1.0000
 39 1   0.0000 1   0.0000 0   0.0000 1   1.0000
 40 1   0.0000 1   0.0000 0   0.0000 1   1.0000
 41 1   0.0000 1   0.0000 0   0.0000 1   1.0000
 42 1   0.0000 1   0.0000 0   0.0000 1   1.0000
 43 1   0.0000 1   0.0000 0   0.0000 1   1.0000
 44 1   0.0000 1   0.0000 0   0.0000 1   1.0000
 45 1   0.0000 1   0.0000 0   0.0000 1   1.0000
 46 1   0.0000 1   0.0000 0   0.0000 1   1.0000
 47 1   0.0000 1   0.0000 0   0.0000 1   1.0000
 48 1   0.0000 1   0.0000 0   0.0000 1   1.0000
 49 1   0.0000 1   0.0000 0   0.0000 1   1.0000
 50 1   0.0000 1   0.0000 0   0.0000 1   1.0000
 51 1   0.0000 1   0.0000 0   0.0000 1   1.0000
  8 1   0.0000 1   0.0000 0   0.0000 1   1.0000
138 1   0.0000 1   0.0000 0   0.0000 1   1.0000
141 1   0.0000 1   0.0000 0   0.0000 1   1.0000
151 1   0.0000 1   0.0000 0   0.0000 1   1.0000
157 1   0.0000 1   0.0000 0   0.0000 1   1.0000
163 1   0.0000 1   0.0000 0   0.0000 1   1.0000
167 1   0.0000 1   0.0000 0   0.0000 1   1.0000
171 1   0.0000 1   0.0000 0   0.0000 1   1.0000
175 1   0.0000 1   0.0000 0   0.0000 1   1.0000
199 1   0.0000 1   0.0000 0   0.0000 1   1.0000
206 1   0.0000 1   0.0000 0   0.0000 1   1.0000
231 1   0.0000 1   0.0000 0   0.0000 1   1.0000
234 1   0.0000 1   0.0000 0   0.0000 1   1.0000
256 1   0.0000 1   0.0000 0   0.0000 1   1.0000
296 1   0.0000 1   0.0000 0   0.0000 1   1.0000
329 1   0.0000 1   0.0000 0   0.0000 1   1.0000
331 1   0.0000 1   0.0000 0   0.0000 1   1.0000
341 1   0.0000 1   0.0000 0   0.0000 1   1.0000
347 1   0.0000 1   0.0000 0   0.0000 1   1.0000
382 1   0.0000 1   0.0000 0   0.0000 1   1.0000
406 1   0.0000 1   0.0000 0   0.0000 1   1.0000
```

```
410 1  0.0000 1  0.0000 0  0.0000 1  1.0000
 52 1  1.0000 1  0.0000 0  0.0000 1  0.7375
 53 1  1.0000 1  0.0000 0  0.0000 1  0.4593
 54 1  1.0000 1  0.0000 0  0.0000 1  0.2172
292 1  1.0000 1  0.0000 0  0.0000 0  1.0000
386 1  1.0000 1  0.0000 0  0.0000 0  1.0000
404 1  1.0000 1  0.0000 0  0.0000 0  1.0000
413 1  1.0000 1  0.0000 0  0.0000 0  1.0000
```

Output file RES.DAT

Calculated results are stored in file RES.DAT and are listed as follow:

```
************************************************************

# Flow over a step for D=1, L=4D STEP =0.4DX0.4D
# CONSTANT VELOCITY AT INLET (unit).
# (phi=0 at Y=0, phi=1 at y=1  )
# for adaptive mesh B
************************************************************


Logical variables:


LNEWT  - flag on momentum diffusion terms:         F
LQUADS - flag for quadrature on diffusion terms:   F
LNCONS - flag on momentum convection terms:        T
LPICON - flag for p initial condition setting:     F
LQNAG  - flag for NAGFEL quadrature:               T
LQUADP - flag for quadrature on pressure-terms:    F
LSECND- flag for ORDER of PROJECTION method:       T
LSCALE- flag for nondimensionalisation:            T
LTMIND- flag for transient b.c.:                   T
LMNITR- flag for monitoring convergence:           T
LSTRIM- flag for stream function calculation:        T
```

```
LSIMPL - flag for SEMI-IMPLICIT T/G:              T
LQUADC - flag for quadrature on convection terms: F
LPLANE - flag for plane or axisymetric version:   T
LELAST - flag for elastic version:                T
LLAST  - flag for last run:                       T
LBODY  - flag for body force:                     F
LREZRL- flag for non-zero Reynolds number:        T
LINERS- flag for linear stream function:          F
LFDBCK- flag for b.cs feedback:                   F
LSAVR - flag for piecewise constant viscosity:    F
LSTGE2- flag to include stage2:                   T
LSTGE3- flag to include stage3:                   T
LDCOPL- flag for decoupled TG:                    F
LFRIZV- flag for freezing velocity field:         F
LFRIZT- flag for freezing stress field:           F


Model specified  :   1 Non-dimen type :  1


Powerlaw index    :  0.75000D+00
Powerlaw const.   :  0.20000D-01
Carreau EZERO     :  0.20000D+00
viscosity1 VISC1 :  0.88880D+00
viscosity2 VISC2 :  0.11120D+00
relax. time EL1   :  0.10000D+01
yield stress      :  0.10000D+01


Density           :  0.10000D+01
Char.velocity     :  0.10000D+01
Char.length       :  0.10000D+01
Viscosity         :  0.10000D+01


Body force vector:  0.00000D+00    0.00000D+00
Reynolds No. 1    :  0.10000D+01
Reynolds No. 2    :  0.10000D+01
extra parm CONST1: -0.88880D+00


Relative time-stepping termination limit :   10000
Tolerance for time-stepping               :  0.10000D-07
```

```
Tolerance for Mass-itn                      :  0.10000D-01
Iterative relaxation factor for Mass-itn :  0.70000D+00

Time-step                    :  0.10000D+00
Crank-N factor               :  0.50000D+00
Alpha                        :  1
extra parm CONST2            :  0.17776D+02
Domain rotation :  0.00000D+00   degrees
Number of elements       :    186
Number of nodes          :    427
Number of boundary nodes :    108
Number of boundary sides :      0
Number of vertex nodes   :    121


Semi-bandwidth (NPBND) :    114

Semi-bandwidth (NQBAND) :    420


Solution at time step :        1, at time :        0.10000


Solution at time step :        2, at time :        0.20000


Solution at time step :        3, at time :        0.30000


Solution at time step :        4, at time :        0.40000


Solution at time step :        5, at time :        0.50000


Solution at time step :        6, at time :        0.60000
```

```
Solution at time step :        7, at time :        0.70000

Solution at time step :        8, at time :        0.80000

Solution at time step :        9, at time :        0.90000

Solution at time step :       10, at time :        1.00000

Solution at time step :       11, at time :        1.10000
   ...
   ...
Solution at time step :     4048, at time :      404.80000

Solution at time step :     4049, at time :      404.90000

Solution at time step :     4050, at time :      405.00000

Solution at time step :     4051, at time :      405.10000

Solution at time step :     4052, at time :      405.20000

Solution at time step :     4053, at time :      405.30000

Solution at time step :     4054, at time :      405.40000
```

Solution at time step :   4055, at time :   405.50000
velocities: stresses:

```
1      0.00000D+00 0.00000D+00
2      0.00000D+00 0.00000D+00
3      0.00000D+00 0.00000D+00
4      0.00000D+00 0.00000D+00
5      0.00000D+00 0.00000D+00
6      0.00000D+00 0.00000D+00
7      0.00000D+00 0.00000D+00
8      0.00000D+00 0.00000D+00

    . . .
    . . .

398    0.10706D+01 0.30667D-01
399    0.10822D+01-0.32587D-01
400    0.10655D+01-0.22193D-01
401    0.10442D+01-0.14815D-01
402    0.00000D+00 0.00000D+00
403    0.90863D-01-0.14356D-01
404    0.10000D+01 0.00000D+00
405    0.96182D+00-0.37778D-01
406    0.00000D+00 0.00000D+00
407    0.12853D+01 0.00000D+00
408    0.13158D+01-0.69328D-01
409    0.71216D+00-0.48842D-01
410    0.00000D+00 0.00000D+00
411    0.86500D+00-0.16385D-01
412    0.10613D+01-0.39244D-01
413    0.10000D+01 0.00000D+00
414    0.10106D+01-0.16502D-02
415    0.10103D+01-0.17406D-01
416    0.10287D+01 0.14446D-01
417    0.17059D+00-0.27052D-01
418    0.19623D+00-0.49700D-02
419    0.12757D+01-0.66698D-01
420    0.66951D+00-0.61399D-01
```

```
421      0.00000D+00 0.00000D+00
422      0.36147D-01-0.12119D-01
423      0.15167D-01-0.50746D-02
424      0.00000D+00 0.00000D+00
425      0.21759D+00 0.00000D+00
426      0.29002D+00-0.44120D-01
427      0.36717D+00-0.49594D-01
pressure:

  1      0.21086729D+01
  2      0.28773410D+01
  3      0.21805443D+01
  4     -0.57484918D+00
  5     -0.79169854D-01
  6      0.00000000D+00
  7      0.87347383D-02
  8      0.21953307D+01
  9      0.21506272D+01
 10      0.19449456D+01
 11      0.19708106D+01
 12      0.15297855D+01
 13      0.27905998D+01
 14      0.40208783D+01
 15     -0.19174742D+01
 16      0.95910445D+00
 17      0.33057012D-01
 18     -0.18466993D+00


        . . .
        . . .


106     -0.10711416D+00
107     -0.20469680D-01
108     -0.84946097D-01
109     -0.82999265D-01
110      0.19400501D+01
111      0.20034016D+01
112     -0.72670093D-01
```

113	-0.80418046D-01
114	0.20820548D+01
115	0.19436175D+01
116	0.21724119D+01
117	0.20303861D+01
118	0.19689844D+01
119	0.20401581D+01
120	-0.62163205D-01
121	-0.26937654D-01

streamfunction:

1	0.00000000D+00
2	0.00000000D+00
3	0.00000000D+00
4	0.00000000D+00
5	0.00000000D+00
6	0.00000000D+00
7	0.10000000D+01
8	0.10000000D+01
9	0.00000000D+00
10	0.00000000D+00
11	0.00000000D+00

. . .
. . .

375	0.44861035D+00
376	0.48386426D+00
377	0.37748368D+00
378	0.21883221D+00
379	0.10000000D+01
380	0.92403542D+00
381	0.78064487D+00
382	0.10000000D+01
383	0.94287803D+00
384	0.50994462D-01
385	0.57993124D-02
386	0.33532125D+00

387	0.23278857D+00
388	0.36402134D+00
389	0.93430241D+00
390	0.81745123D+00
391	0.45188358D+00
392	0.35898217D+00
393	0.28233318D+00
394	0.16500533D+00
395	0.25455891D+00
396	0.26059423D+00
397	0.22330041D+00
398	0.38682069D+00
399	0.66023669D+00
400	0.67661023D+00
401	0.53832829D+00
402	0.00000000D+00
403	0.77095886D-01
404	0.91133456D+00
405	0.91093793D+00
406	0.10000000D+01
407	0.10000000D+01
408	0.73221092D+00
409	0.61671791D+00
410	0.10000000D+01
411	0.92995473D+00
412	0.77753320D+00
413	0.60245174D+00
414	0.50715973D+00
415	0.65400850D+00
416	0.41063662D+00
417	0.24919917D+00
418	0.45131421D+00
419	0.50242952D+00
420	0.26192597D+00
421	0.00000000D+00
422	0.90949611D-01
423	0.51893217D-01
424	0.00000000D+00

425	0.10000000D+01
426	0.14950184D+00
427	0.25706090D+00

Subject Index